Elevating

React Web Development

with **Gatsby**

Practical guide to building performant, accessible, and interactive web apps with React and Gatsby.js 4

Samuel Larsen-Disney

BIRMINGHAM—MUMBAI

Elevating React Web Development with Gatsby

Group Product Manager: Pavan Ramchandani
Publishing Product Manager: Ashitosh Gupta
Senior Editor: Hayden Edwards
Content Development Editor: Rashi Dubey
Technical Editor: Joseph Aloocaran
Copy Editor: Safis Editing
Project Coordinator: Rashika Ba
Proofreader: Safis Editing
Indexer: Hemangini Bari
Production Designer: Roshan Kawale
Marketing Coordinator: Anamika Singh

First published: January 2022
Production reference: 1190122

Published by Packt Publishing Ltd.
Livery Place
35 Livery Street
Birmingham
B3 2PB, UK.

ISBN 978-1-80020-909-1
www.packt.com

Contributors

About the author

Sam studied computer science at King's College, London. He helped design and build American Express' websites. He then moved to BehaviourLab where he led frontend development, before deciding he wanted to get out of finance. He has since become a senior frontend engineer at Zone. He is most at home coding in React, JavaScript, GraphQL, and Gatsby but is always open to learning something new. When coding, he likes to ensure his code is accessible and performant. In the last year, Sam has contributed 1,300+ times to open source projects. He enjoys teaching the next generation to code through his articles and presentations, and at hackathons.

> *My passion for web development is the direct result of some awesome engineers who took the time to teach me their craft. I would particularly like to thank Adam Wilkinson and Diego Abizaid who challenged what I thought was possible in the browser. I would also like to thank Yannis Panagis, Ryan Gregory, Çelik Köseoğlu, Meghan Avery, Arthur Ceccotti, Ruben Casas, and Joshua Gabrel, who all made my web development journey unique.*

About the reviewer

Benjamin Read has been developing websites for the past decade. That covers a lot of ground: from the advent of the iPhone and responsive design to today's isomorphic, serverless, and reactive web applications. A few years ago, he came across an exciting new project called Gatsby. Gatsby catapulted his interests in static sites, GraphQL, content management systems, and other concepts that continue to propel his technical interests today. When he's not working, contributing to open source, or tinkering with new technology, Ben is usually found spending time with his wife and three children, or reading a good book.

Table of Contents

Part 2: Going Live

6

Improving Your Site's Search Engine Optimization

7

Testing and Auditing Your Site

8

Web Analytics and Performance Monitoring

9
Deployment and Hosting

Part 3: Advanced Concepts

10
Creating Gatsby Plugins

11
Creating Authenticated Experiences

12
Using Real-Time Data

13
Internationalization and Localization

Index

Other Books You May Enjoy

Preface

Gatsby is a powerful React static site generator that enables you to create lightning-fast web experiences. With this latest version of Gatsby, you can combine your static content with server-side rendered and deferred static content to create a fully rounded application. Elevating React Web Development with Gatsby provides a comprehensive introduction for anyone new to GatsbyJS and will have you up to speed in no time.

Complete with hands-on tutorials and projects, this easy-to-follow guide starts by teaching you the core concepts of GatsbyJS. You'll then discover how to build performant, accessible, and scalable websites by harnessing the power of the GatsbyJS framework. This book takes a practical approach to help you to build anything from your personal website through to large-scale applications with authentication and make your site rise through those SEO rankings.

By the end of this book, you will know how to build client websites your users will love. Every aspect of performance and accessibility is a point of emphasis with this tool and you will learn how to squeeze every ounce of benefit out of it through the book's material.

Who this book is for

This book is for web developers who want to use GatsbyJS with React to build better static and dynamic web apps. Prior experience of React basics is necessary. Basic experience of Node.js will help you to get the most out of this book.

What this book covers

Chapter 1, An Overview of Gatsby.js for the Uninitiated, provides baseline knowledge of what Gatsby.js is and explains the guiding principles we will be using in later chapters to build our web application.

Chapter 2, Styling Choices and Creating Reusable Layouts, shows how to make an informed choice about the way you would like to style your application. We will cover using CSS, SCSS, styled-components, and Tailwind.css.

Chapter 3, *Sourcing and Querying Data (from Anywhere!)*, gets you to a position where you can comfortably source and ingest data into your Gatsby projects from a multitude of different sources.

Chapter 4, *Creating Reusable Templates*, explains how to use your sourced data to programmatically create site pages, blog posts, and more!

Chapter 5, *Working with Images*, shows you how to master the art of adding responsive images to your Gatsby site without impacting performance.

Chapter 6, *Improving Your Site's Search Engine Optimization*, explains how SEO works, what search engines look for within your site pages, and how to improve your site's presence on the web.

Chapter 7, *Testing and Auditing Your Site*, covers testing and auditing your application using industry-standard tooling.

Chapter 8, *Web Analytics and Performance Monitoring*, explains how to add analytics to your site and use your audience to make your site even better!

Chapter 9, *Deployment and Hosting*, shows how to take the project we have been working on and deploy it for the world to see!

Chapter 10, *Creating Gatsby Plugins*, covers creating source and theme plugins and explains how to contribute them to the Gatsby plugin ecosystem.

Chapter 11, *Creating Authenticated Experiences*, shows you how to add protected routes to create logged-in experiences on your site.

Chapter 12, *Using Real-Time Data*, explains how you can use sockets to create experiences that make use of real-time data.

Chapter 13, *Internationalization and Localization*, covers patterns you can use to make translating your site as it scales simple.

To get the most out of this book

All code examples have been tested using Gatsby 4.4.0 on macOS. However, they should work with future 4.x releases too.

Software/hardware covered in the book	Operating system requirements
Gatsby 4.x	Windows, macOS, or Linux
React 17.x	Windows, macOS, or Linux
Node.js LTS	Windows, macOS, or Linux

This book assumes you have an **Integrated Development Environment** (IDE) *installed that you are comfortable using.*

If you are using the digital version of this book, we advise you to type the code yourself or access the code from the book's GitHub repository (a link is available in the next section). Doing so will help you avoid any potential errors related to the copying and pasting of code.

Download the example code files

You can download the example code files for this book from GitHub at `https://github.com/PacktPublishing/Elevating-React-Web-Development-with-Gatsby-4`. If there's an update to the code, it will be updated in the GitHub repository.

We also have other code bundles from our rich catalog of books and videos available at `https://github.com/PacktPublishing/`. Check them out!

Download the color images

We also provide a PDF file that has color images of the screenshots and diagrams used in this book. You can download it here: `https://static.packt-cdn.com/downloads/9781800209091_ColorImages.pdf`.

Conventions used

There are a number of text conventions used throughout this book.

`Code in text`: Indicates code words in text, database table names, folder names, filenames, file extensions, pathnames, dummy URLs, user input, and Twitter handles. Here is an example: "Create a `gatsby-config.js` file in your root directory and add the following."

A block of code is set as follows:

```
module.exports = {
  plugins: [],
};
```

When we wish to draw your attention to a particular part of a code block, the relevant lines or items are set in bold:

```
import React from "react"
import {Link} from "gatsby"

export default function Index() => {
    return (
        <div>
            <h1>My Landing Page</h1>
            <p>This is my landing page.</p>
            <Link to="/about">About Me</Link>
        </div>
    )
}
```

Any command-line input or output is written as follows:

```
gatsby develop -H 0.0.0.0
```

Bold: Indicates a new term, an important word, or words that you see onscreen. For instance, words in menus or dialog boxes appear in **bold**. Here is an example: "When you hit the **Play** button above the query, you will see the result of that query on the central right column, with a JSON object containing the data property and our query's result inside it."

> **Tips or Important Notes**
> Appear like this.

Get in touch

Feedback from our readers is always welcome.

General feedback: If you have questions about any aspect of this book, email us at customercare@packtpub.com and mention the book title in the subject of your message.

Errata: Although we have taken every care to ensure the accuracy of our content, mistakes do happen. If you have found a mistake in this book, we would be grateful if you would report this to us. Please visit www.packtpub.com/support/errata and fill in the form.

Piracy: If you come across any illegal copies of our works in any form on the internet, we would be grateful if you would provide us with the location address or website name. Please contact us at copyright@packt.com with a link to the material.

If you are interested in becoming an author: If there is a topic that you have expertise in and you are interested in either writing or contributing to a book, please visit authors.packtpub.com.

Part 1: Getting Started

Upon finishing this part, you should have a clear understanding of what Gatsby.js is. You should also be at a point where you can comfortably develop basic sites in Gatsby.js on your local machine.

In this part, we include the following chapters:

- *Chapter 1, An Overview of Gatsby.js for the Uninitiated*
- *Chapter 2, Styling Choices and Creating Reusable Layouts*
- *Chapter 3, Sourcing and Querying Data (from Anywhere!)*
- *Chapter 4, Creating Reusable Templates*
- *Chapter 5, Working with Images*

1

An Overview of Gatsby.js for the Uninitiated

In this book, we will take your existing React knowledge and supplement it with Gatsby.js (which we will refer to as Gatsby from now on) to create performant and accessible static sites. I hope to give you the tools you need to create better websites using Gatsby and get you to join the static site revolution. So, *happy hacking!*

This chapter starts with a brief historical look at the static web and why Gatsby was created. Then, we'll think about what Gatsby is and how it builds on React. Next, we'll go through some of the use cases of Gatsby and identify Gatsby's competitors. Finally, we'll set up a basic Gatsby project, having created our first few pages.

In this chapter, we will cover the following topics:

- A brief history of the static web
- What is Gatsby?
- Gatsby use cases
- Gatsby's competitors
- Setting up a project

Technical requirements

The code present in this chapter can be found at `https://github.com/PacktPublishing/Elevating-React-Web-Development-with-Gatsby-4/tree/main/Chapter01`.

A brief history of the static web

Static sites have been around nearly as long as the internet itself. They are the original blueprint for any website – **HyperText Markup Language** (**HTML**), **Cascading Style Sheets** (**CSS**), and **JavaScript** (**JS**). In the 1990s, HTML was the only publishing mechanism for the web. To get content on the internet, you would have to create a static HTML file and expose it to the internet via a server. If you wanted to modify one of your web pages, you would need to change its corresponding HTML file directly.

While learning HTML is part of primary education these days, back in the 1990s, it was a novel skill to understand and write the language. Creating or editing content was costly, as you would require someone with this skill set for every modification. Luckily, **Content Management Systems** (**CMSes**) (WordPress, Drupal, and so on) soon swooped in to allow non-technical users to control a webpage's design and content. It also gave users the ability to store and manage files via a user interface. CMSs continue to be utilized today with increasing popularity. The number of websites using a CMS has risen from 23.6% to 63% in the last decade. Over 75 million sites use WordPress today – that's 30% of the web!

At an almost identical pace, frontend frameworks and libraries have gained notoriety. Building single-page applications became commonplace. Today, the most dominant UI library in the JS world is Facebook's React.js, which is a small library with a handful of functions but some big ideas – a virtual DOM, **JavaScript Syntax Extension** (**JSX**), and componentization. There is no denying how much impact React has had on web development. In 2020, 80% of JS developers had used it, and 70% of JS developers said they would use it again.

Frontend frameworks have entirely changed how developers approach web development, giving them the flexibility to focus on functionality over content and drastically speeding up their workflows. But you're only as fast as your slowest team member. The clunky nature of CMS platforms was revealed when developers started to employ these frameworks and integrate them with CMSs. Traditional CMS workflows made use of databases and environments that frontend frameworks had removed from the equation. Combining this with CMS security and bottleneck issues led to the rebirth of static sites.

Kyle Mathews, the founder of Gatsby, was a catalyst for this trend. He noticed that the expectations on website accessibility and performance increased dramatically. He observed apps investing millions of dollars in user experience. There is no denying that the disparity between a 2005 and 2015 website was significant. In a competitive environment such as the web, you have to have a product that can stand out. Mathews took a step back, identified gaps in existing tooling, and asked what the ideal product might be. This research is what led him to create Gatsby.

It's almost poetic that we have gone full circle and returned to static content because there is no beating it when it comes to speed and performance.

What is Gatsby?

Gatsby is a free, open source static site generator that harnesses React. Static site generators are software applications that create static pages from a template or component and supplement them with content from a source. Static site generators are an alternative to a more traditional database-driven CMS, such as WordPress. In these conventional systems, content is managed and stored in a database. When the server receives a particular URL request, the server retrieves data from the database, mixes it with a template file, and generates an HTML page as its response. Generating HTML on demand can be a time-consuming process and can leave the user twiddling their thumbs or, worse, leaving your site. Bounce rates (the percentage of visitors to a particular website who navigate away from the site after viewing only one page) hover below 10% for websites that take less than 3 seconds to load, but the number jumps to 24% for a 4-second load time and 38% for a 5-second load time.

Static site generators like Gatsby, on the other hand, generate pages during a build process. During this process, Gatsby brings in data to its GraphQL layer, where it can be queried in pages and templates. The requested data is then stored in JSON and accessed by the built pages, which are composed of HTML, JS, and CSS files. A user can deploy these generated pages to a server. When it receives a request, the server responds with predetermined, static, rendered HTML. As these static pages are generated at build time, they eliminate the latency that databases would introduce. You can even do away with web servers altogether and have your site served via a CDN pointing to a storage medium, such as an AWS **Simple Storage Service (S3)** bucket. The difference is striking; web experiences built with Gatsby are lightning fast, as nothing can be faster than sending static content.

> **Important Note**
>
> A static site can contain dynamic and exciting experiences! It is a common misconception that "static" means the site is stationary. This could not be further from the truth. The word "static" only refers to the manner in which files are retrieved by a client.

While Gatsby is known for static site generation, recent versions also include server-side and deferred static generation, rendering functionality for when static generation is not enough.

Aside from creating a blazing-fast user experience, Gatsby also has a focus on developer experience. As we learn and build, I'm sure you will start to recognize how easy it is to use. The way it achieves this can be broken down into four steps.

Community

Gatsby has an incredibly supportive community backing. At the time of writing, over 3,600 people have contributed to the Gatsby repository. This is further amplified by the plugin ecosystem surrounding Gatsby; the community has created more than 2,000+ plugins that abstract complex functionality that other developers may wish to use in their own projects. These plugins are distributed as packages stored on a JS repository, such as **NPM**, that can be added to your project in a few lines. They can extend your site by sourcing content, transforming data, creating pages, or theming your application.

Sourcing content from anywhere

Every day, the amount of data we need to combine to create experiences is rising. In traditional React applications, managing multiple sources of data could become a nightmare. Storing, massaging, merging, and querying data all require complex solutions that struggle to scale.

Gatsby does this differently. Whether you are sourcing data from a CMS, real-time database, or even a custom **Application Programming Interface (API)**, you can merge all of this data into a unified data layer. The Gatsby community is constantly contributing source plugins to allow you to ingest data from your favorite sources with ease. Nine times out of ten, you won't need to write a single line of code to source your data, but for the times when you do, we will be covering plugin creation in *Chapter 10, Creating Gatsby Plugins.*

Once ingested into this data layer, we can explore and query all our sources of data in one place using a uniform data layer. Using the power of GraphQL, we can query our data in the same way when rendering pages regardless of their source. The GraphQL layer is transitory and doesn't exist after the application has been built, so doesn't affect the size of your production site. If GraphQL is something new to you, don't worry – I will be explaining how it works in *Chapter 3, Sourcing and Querying Data (from Anywhere!).*

Building tooling you already know

Often when we approach new technologies, we are faced with a steep learning curve as we understand new syntax and ways of thinking. In Gatsby, we build on your existing knowledge of React instead of starting from scratch. Underpinning all of our code is the same React component model many of you already know. You should feel pretty confident from the beginning, as the code should look familiar, and if you're not, Gatsby can also help you learn React from a more "content-driven" approach.

Supercharging web performance

As web developers, we can spend considerable time tinkering with websites to squeeze every ounce out of their performance. Sometimes, this can take as long, if not longer, than building the design. Also, performance gains can sometimes be undone instantly by a change to the site design outside of your control. It's because of this that some large organizations have dedicated teams to improve site performance. But it doesn't have to be this way! As we start to build together, you will see that load times go from seconds to milliseconds, and your site will feel far more responsive than a conventional React app. Gatsby has plenty of tricks up its sleeve that improve performance, some of which we will touch on at the end of this chapter. It also turns your site into a **Progressive Web App** (**PWA**) with just a few lines of code – if that's not cool, I don't know what is!

> **Important Note**
> An essential distinction between Gatsby and React is that Gatsby is a "framework," not a "library." When using a library, you control your application flow; you call it when you need it. When using a framework, however, there is an inversion of control. Frameworks command that you adhere to a particular flow and layout defined by them. Working within a framework can often be seen as a benefit, as any developer familiar with the framework will know where to find relevant files and code.

I hope you are beginning to see some of the great reasons why Gatsby is such a powerful tool. Let's now see it in action.

Gatsby use cases

You might be starting to realize that Gatsby could have applications across many different kinds of websites. Since Gatsby's v1 launch in 2017, the framework has been used in a multitude of different ways by companies both big and small. Here, I want to highlight some examples of use cases where Gatsby excels and suggest why companies may have chosen Gatsby for these sites.

> **Tip**
> While reading about these example sites here is great, I highly encourage you to visit them via your own device. One of Gatsby's best features is the speed of the sites it creates, and it is essential to experience this for yourself to understand the benefit.

Documentation sites

Documentation sites are a perfect use case for Gatsby as their content is primarily, if not entirely, static. Their content does not shift often either, with pages needing infrequent updates. Their static nature means that we can generate all page routes during the build process and load them onto a CDN, meaning that when a page is requested, the request is near-instant. It is for this reason that you see sites such as the official React documentation (`https://reactjs.org`) being made with Gatsby:

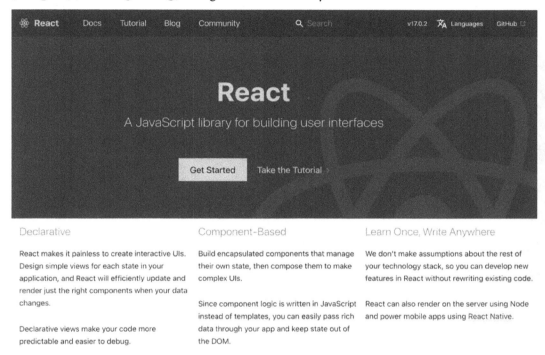

Figure 1.1 – The React documentation website

Due to the infrequent nature of updates to documentation pages, you can automate the build and deployment of your site as and when changes to documentation are made. With GitHub integrations or webhooks, you can get your documentation site to redeploy each change to a master branch or on a daily basis, for example. We will be exploring how to create these kinds of processes in *Chapter 9, Deployment and Hosting*.

Online courses

Online courses often have a unique structure – the majority of their content is in static learning modules, but they also require a small quantity of authenticated routes for logged-in user experiences.

Websites such as *DesignCode.io* (`https://designcode.io/courses`) utilize Gatsby for their static content, meaning their static pages are incredibly performant, and they then render authenticated routes on the client. While this does increase bundle size, as they need to ship more JS, the benefit of the fast static pages far outweighs the cost of heavier authenticated pages:

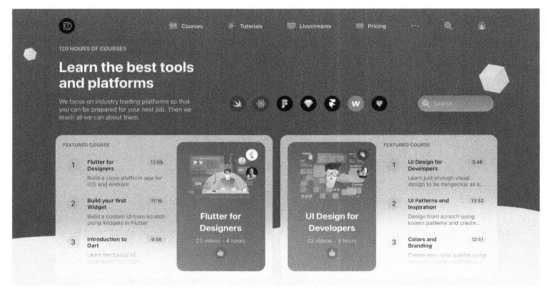

Figure 1.2 – The DesignCode.io website

One of the most popular sources of data for Gatsby is MDX. MDX is a powerful format that allows you to write JSX within Markdown. Why is it awesome? Because you can include React components alongside documentation with no hassle at all. React components can be far more interactive and dynamic than text, and as a result, it is a powerful format to create online courses on, as you can create content that is more enticing for the user. Perhaps a more interactive course is a more memorable one? We will be diving into MDX in detail in *Chapter 3, Sourcing and Querying Data (from Anywhere!)*.

SaaS products

When selling **Software as a Service (SaaS)** online, your website's performance can be considered a reflection of your product's performance. As a result, having a clunky website can be the difference between your product being a success or not. As mentioned previously, this is an example where you could go down a rabbit hole to improve your site's performance. Companies such as *Skupos* (`https://www.skupos.com/`) use Gatsby to get more performance benefits for free. Gatsby also works wonders for **Search Engine Optimization (SEO)**. As pages are prerendered, all your page content is available to web crawlers such as Googlebot to navigate to your site. The speed and SEO improvements help their product's website stand out and give the user confidence that they know what they are doing when it comes to technology:

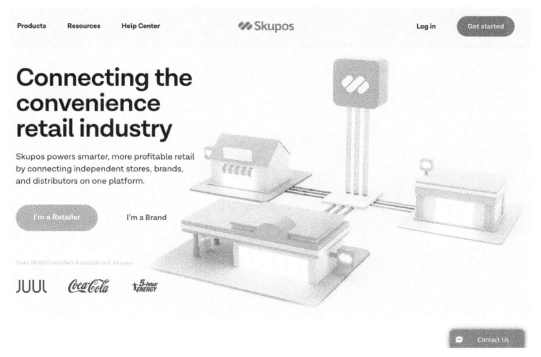

Figure 1.3 – The Skupos website

Skupos also supplement their site pages with metadata and alt-text, which further aids web crawlers in understanding site content. The more web crawlers understand your site's content, the better your search engine ranking will be.

Design agencies and photo-heavy sites

In cases where your work is more visual, your site often needs to make use of large quantities of high-resolution images. We've all visited a website and felt like we were transported back to the dial-up days as we've waited for large image files to load. This common mistake is often amplified further by a large amount of cumulative layout shift that happens when loading images. Gracefully handling the image's loading state to avoid this can be a headache.

Gatsby performs magic for images within its application. It utilizes the `sharp` library (`https://github.com/lovell/sharp`) under the hood to convert your large images into smaller web-friendly sizes. When your website loads, it will first load in a smaller resolution version before blurring up to the maximum resolution required. This results in no layout shift and a far less "jumpy" experience for your site visitor. A great example of this is on the *Call Bruno Creative Agency* (`https://www.callbruno.com/en/reelevant`) website developed with Gatsby:

Figure 1.4 – The Call Bruno Creative Agency website

They use lots of imagery across their project pages, but the image load does not take you out of the experience. We will get into detail on handling images in *Chapter 5, Working with Images.*

By exploring these sites, we can see examples across industries where Gatsby is helping companies get ahead of their competition.

Gatsby's competitors

While this book focuses on Gatsby, it is crucial to understand that it is not the only React static site generator on the market. The competitor most often uttered in the same breath is Next.js.

Until recently, the key difference between Next.js and Gatsby was server-side rendering. Like Gatsby, a Next.js application can be hosted statically, but it also used to be able to server render pages where Gatsby could not. Instead of deploying a static build, a server is deployed to handle requests. When a page is requested, the server builds that page and caches it before sending it to the user. This means that subsequent requests to the resource are faster than the first call. As of version 4, Gatsby can have all of its pages prebuilt statically or it can create a hybrid build – a mixture of static and server-side rendered content. We will discuss this more in *Chapter 9, Deployment and Hosting.*

One major drawback to Next.js is its data security. When building Gatsby sites as static builds, data is only taken from the source at build time, and as the content is static, it is secure. Next.js keeps data stored on the server and, as such, it is easier to exploit. Next.js commonly requires more initialization if you wish to set it up via a server or using databases. This also means that there is more maintenance required in Next.js applications. Both Next.js and Gatsby have additional utilities to help with the handling of images. Gatsby, however, can make images more performant on statically rendered pages, while Next cannot.

The good news is that all static site generators follow a similar process. *The skills and mentality you learn in this book are easily transferable to a different generator in the future should you decide you want to make the switch.*

Now that we understand where Gatsby excels, let's start creating our first Gatsby project.

Setting up a project

In order to help you put into practice what you're learning, we will be building a project together. Throughout this book, we will be working to build a personal portfolio, something that every developer needs and therefore something I think will be relevant for most readers. The portfolio will contain blog pages to aid your learning in public, project pages to demonstrate your work, a stats page showcasing interesting metrics on your site, as well as many more features that will help your portfolio stand out from the crowd.

Throughout this book, you will be faced with options. We will discuss different implementations for styling your site, as well as data sources you may want to implement. This should give you the flexibility to align it with your current knowledge. Alternatively, you can throw yourself in the deep end – the choice is up to you. Everywhere there is a choice, I will also provide my personal recommendation for what might be best if you can't decide.

To see a finished version of the portfolio we will be building, visit this link:

`https://elevating-react-with-gatsby.sld.codes/`

> **Tip**
>
> Refer to the code repository (`https://github.com/PacktPublishing/Elevating-React-Web-Development-with-Gatsby-4`) that accompanies this book if you're struggling at any point. It includes a copy of the project as it should appear after every chapter.

To start using Gatsby, we need to ensure we have a few prerequisite tools set up on our machines. Most of these prerequisites are most likely already on your device if you are a React developer, although I would still encourage you to read through this list, as some of your tools may need an update.

Node.js version 14.15.0+

As of version 4.0, Gatsby supports all Node.js versions greater than 14.15.0. You can quickly check if you have Node.js installed by opening up a terminal window and typing the following:

```
node -v
```

If you have Node.js installed, this should print a version number. However, if you receive an error, you can download Node.js by navigating to the Node.js website (`https://nodejs.org`). Node.js comes bundled with npm, a package repository, package manager, and command-line tool that we will be using to install Gatsby.

> **Tip**
> You're most likely already using Node.js, and some of your pre-existing projects may require a different version than the requirements specified here. If you need to manage multiple versions of Node.js on the same device, you should check out the **Node.js Version Manager** (**NVM**)(`https://github.com/nvm-sh/nvm`). It gives you access to valuable commands, including installing new versions and switching between minor and major versions of Node.js.

Gatsby command-line interface

The Gatsby **Command-Line Interface** (**CLI**) is a tool built by the core Gatsby team; it allows you to perform standard functions, such as creating new Gatsby projects, setting up local development servers, and building your production site. Although you can use it on a per-project basis, it is far more common to install the CLI globally so that you can use its features across multiple Gatsby projects without having to install it as a package in each project – got to save that hard-drive space!

To install the CLI globally, `npm install` it with the global flag:

```
npm i -g gatsby-cli
```

To verify its installation, open up a terminal window and type the following:

```
gatsby --help
```

If running this provides a list of commands and does not error out, then you're good to go.

> **Important Note**
> Throughout this book, I use npm as my package manager. If you prefer Yarn, you can use the Yarn equivalent commands.

Directory and package setup

Here, we will begin to create the files and folders we need to start our project, as well as install necessary dependencies such as React and Gatsby.

First, create a folder to house our project. You can call it whatever you like. Throughout this book, I will refer to this folder as the `root` folder of the application. Open a terminal and navigate to your `root` folder. Initialize a new package in this folder by running the following:

```
npm init -y
```

With the package now initialized, let's install React and Gatsby:

```
npm i gatsby react react-dom
```

Open your `root` folder in your favorite **Integrated Development Environment (IDE)**. You should notice that it now contains three new items, `package.json`, `package-lock.json`, and a `node-modules` folder. Opening your `package.json`, you should see the following:

```
{
  "name": "gatsby-site",
  "version": "1.0.0",
  "description": "",
  "main": "index.js",
  "scripts": {
    "test": "echo \"Error: no test specified\" && exit 1"
  },
  "keywords": [],
  "author": "",
  "license": "ISC",
  "dependencies": {
    "gatsby": "^4.4.0",
    "react": "^17.0.2",
    "react-dom": "^17.0.2"
  }
}
```

In the preceding example, you can see that this file now contains references to the dependencies we have just installed.

Development scripts

Let's start by modifying `package.json` so that it contains some useful scripts that will speed up our development process:

```
{
  "name": "gatsby-site",
  "version": "1.0.0",
  "description": "",
  "main": "index.js",
```

```
"scripts": {
  "build": "gatsby build",
  "develop": "gatsby develop",
  "start": "npm run develop",
  "serve": "gatsby serve",
  "clean": "gatsby clean"
},
"keywords": [],
"author": "",
"license": "ISC",
"dependencies": {
  "gatsby": "^4.4.0",
  "react": "^17.0.2",
  "react-dom": "^17.0.2"
}
}
```

Let's break down these scripts:

- build: Runs the Gatsby CLI's build command. This creates a compiled, production-ready build of our site. We will learn more about this in *Chapter 9, Deployment and Hosting*.

- develop: Runs the Gatsby CLI's develop command. We will review it in detail in the next section, *Creating your first few pages*.

- start: The start script redirects to the develop script. This is in place as it is common to start packages with a start script.

- serve: Runs the Gatsby CLI's serve command to serve up a Gatsby build folder. This is a useful way to review a production build.

- clean: The clean script utilizes the Gatsby CLI's clean command. This deletes the local Gatsby cache and any build data. It will be rebuilt with the next develop or build command.

All of these scripts can be run from the root folder with the following command:

```
npm run script-name
```

Simply replace script-name with the name of the script you would like to run.

You'll notice the absence of a test script. Don't worry – we will get into how to test a Gatsby application in *Chapter 7, Testing and Auditing Your Site.*

Framework files and folders

As mentioned, Gatsby is a framework. Frameworks require certain files to exist in order to work. Let's set up our project with the files and folders where Gatsby expects to find them.

Create a `gatsby-config.js` file in your `root` directory and add the following:

```
module.exports = {
  plugins: [],
};
```

As the name might suggest, the `gatsby-config.js` file is the core configuration file for Gatsby. We will be coming back to this file frequently as we build out our project. By the time we are done with it, it will be full of plugins, metadata, styling, and even offline support.

Create `gatsby-browser.js` and `gatsby-node.js` files in your `root` directory. Both of these files can be left blank for now. The `gatsby-browser.js` file contains any code we would like to run on the client's browser. In the next chapter, we will be using this file to add styles to our website. The `gatsby-node.js` file contains code we would like to run during the process of building our site.

Finally, create an `src` folder in your `root` directory. This folder will contain the majority of our development work, much like in a traditional React application. Pages we create and components we define will all be contained within this folder.

Before we go any further, let's make sure we have our version control tracking the right files.

Using version control

I suspect many of you would like to use version control while you build out your Gatsby site. To ensure Git tracks only the files that matter, create a `.gitignore` file and add the following:

```
node_modules/
.cache/
public
```

These lines stop our dependencies, Gatsby builds, and cache folders from being tracked.

Creating your first few pages

We now have all the underlying code we need set up to allow us to start creating pages. In this section, we will create a three-page website using Gatsby. It's important to note that this is a basic example purely designed to solidify your understanding of how Gatsby works before we worry about styling and additional functionality.

Navigate to your `src` directory and create a new folder called `pages`. Any JS files we create within the `pages` folder will be treated as a route by Gatsby. This also applies to subfolders within the `pages` folder. There is, however, one exception – files called `index.js` are treated as the root of their directory. Let's make sense of this with a few examples:

- `src/pages/index.js` will map to `yourwebsite.com`.

- `src/pages/about.js` will map to `yourwebsite.com/about`.

- `src/pages/blog/my-first-post.js` will map to `yourwebsite.com/docs/my-first-post`. While we won't be setting up a page at this URL now, we will start using routes such as this one in *Chapter 3, Sourcing and Querying Data (from Anywhere!)*.

- `src/pages/404.js` will map to any page that does not resolve on `yourwebsite.com`.

> **Important Note**
>
> Any React components you place in the `pages` folder will become navigable routes on your site. As such, it is best to separate your components from your pages. A common pattern is to create a `components` folder that sits next to your `pages` folder in the `src` directory and import components you want to use in your pages.

The index page

Create an `index.js` file in your `pages` folder. As the index of the `pages` folder, this will become the landing of your website. We can now populate this file with the following code:

```
import React from "react"

const Index = () => {
    return (
        <div>
```

```
            <h1>My Landing Page</h1>
            <p>This is my landing page.</p>
        </div>
    )
}

export default Index
```

The contents of this file should look familiar; it's just a simple stateless ReactJS component.

We could have also defined it as:

```
import React from "react"

export default function Index(){
    return (
        <div>
            <h1>My Landing Page</h1>
            <p>This is my landing page.</p>
        </div>
    )
}
```

Both examples will output the exact same result, so it's just personal preference.

The about page

In a similar fashion, we can create an about page. Here, you have a choice – you can either create this page at src/pages/about.js or at src/pages/about/index.js. The question I always ask myself when deciding which option to go with is whether the page will have sub-pages. In the case of an about page, I think it's unlikely to contain any sub-pages, so I will opt for src/pages/about.js:

```
import React from "react"

export default function About(){
    return (
        <div>
            <h1>My About Page</h1>
            <p>This is a sentence about me.</p>
```

```
        </div>
    )
}
```

Here, we have defined another simple React component containing a heading and paragraph to create our about page.

The 404 page

Gatsby expects to find a 404.js file in your pages directory. This page is special. It contains the page that will be shown when Gatsby cannot find a page that was requested. I am sure you have come across "Page not found" pages before. Without this page, on requesting a non-existent route, the browser will not find any resource and show a browser error to the user. While the 404 page is another form of displaying the same error, by creating this page, we can manage the error ourselves. We can link to working pages on our site or even suggest the page they might have been trying to visit (more on this in *Chapter 3, Sourcing and Querying Data (from Anywhere!)*).

Let's create our 404 page now in src/pages/404.js:

```
import React from "react"

export default function NotFound() {
    return (
        <div>
            <h1>Oh no!</h1>
            <p>The page you were looking for does not
                exist.</p>
        </div>
    )
}
```

You should be starting to see a pattern. Creating pages is as simple as defining React components – something you should be familiar with already.

Trying the develop command

At this point, you've actually already created a fully working website. Congratulations! To test it out, open a terminal at your root directory and run the following:

```
npm run start
```

As you will recall from our `package.json`, this will run the `gatsby develop` command. This will take a few seconds to run, but you should then see some terminal output that looks like this:

```
You can now view gatsby-site in the browser.
  http://localhost:8000/
```

You can now open a browser of your choice and navigate to `http://localhost:8000/`, and you should be greeted with something like this:

My Landing Page

This is my landing page.

About me

Figure 1.5 – The landing page preview

This is the rendered version of our `index.js` page component. You can modify the URL in your browser to `http://localhost:8000/about` to see your `about` page and `http://localhost:8000/404` to see your `404` page. You can also see your `404` page in development by navigating to any invalid route and pressing the **Preview custom 404 page** button.

Tip

If you don't want to manually navigate to the browser and type in the URL, you can modify our scripts by appending the `gatsby develop` command with the `-o` option. This instructs Gatsby to open your default browser and navigate to the site automatically when you run the `develop` command.

---

gatsby develop in detail

Running `gatsby develop` starts the Gatsby development server. This might be a little confusing, as we have previously mentioned how a Gatsby site is delivered as static content, but it's actually there to speed up your development process.

Imagine your site contains 10,000 pages; building the entirety of your site every time you make a small change to one page would take a long time. To get around this in development, Gatsby uses a Node.js server to build only what you need as and when it's requested. Due to it building on demand, it can negatively affect the performance of a page and *you should never test performance on a page in development for this reason.*

Once the server is up, you can continue to edit your code without rerunning the command. The development server supports hot reloading, a concept that should be familiar to you.

The `develop` command has a number of built-in options that allow you to customize it:

- `-H, --host`: Allows you to modify the host
- `-p, --port`: Allows you to modify the port Gatsby runs on
- `-o, --open`: Opens your project in the browser
- `-S, --https`: Turns on HTTPS

You can view your site on any device connected to the same network by using the host option. This can be useful when you want to compare how your site behaves on mobile browsers with that of a desktop experience. To achieve this, run the following command:

```
gatsby develop -H 0.0.0.0
```

If the command is successful, you will see a subtle difference in the output:

```
You can now view gatsby-site in the browser.
  Local:            http://localhost:8000/
  On Your Network:  http://192.168.1.14:8000/
```

The `develop` command has added a URL for testing on your network. Typing this into a browser on any device connected to the same network will render your site.

Connecting your pages

Now that you have multiple pages, you may want to navigate between them. There are two different ways of achieving this – with the Gatsby Link component or via programmatic navigation. To some of you, these components and functions may sound familiar; this is because Gatsby wraps the `reach-router` (`https://reach.tech/router`) library for navigation. For those who haven't used `reach-router` before, the library comes with support for server-side rendering and routing accessibility functionality built in. Gatsby has built on and enhanced this functionality to meet its high standards for user accessibility, ensuring a great website experience regardless of who you are.

The Gatsby Link component

It's important to use the Gatsby `<Link/>` component as a replacement for the `<a/>` tag whenever you are linking to a page that is internal. The `<Link/>` component works just like an `<a/>` tag, with one important distinction – it enables prefetching. Prefetching is the act of loading a resource before it is required. This means that when the resource is requested, the time waiting for that resource is decreased. By prefetching the links on your page, your next click navigates to content that is already loaded and is therefore practically instant. This is particularly noticeable on mobile devices in areas with reduced network conditions that would normally have a delay when loading pages.

The first place you could add a `Link` component is to your `404` page. It's common for these pages to have a button that says something like "Take me home" that, when clicked, navigates to the landing page:

```
import React from "react"
import {Link} from "gatsby"

export default function NotFound(){
    return (
        <div>
            <h1>Oh no!</h1>
            <p>The page you were looking for does not
                exist.</p>
            <Link to="/">Take me home</Link>
        </div>
    )
}
```

As you can see in the preceding code block, the `Link` component has a prop called `to`; this needs to be passed to the page that you want to navigate to relative to the root of your website. By passing the `"/"` prop, Gatsby will navigate to the root of your website.

You can also add a link to the `about` page from the `index` page:

```
import React from "react"
import {Link} from "gatsby"

export default function Index() => {
    return (
        <div>
```

```
        <h1>My Landing Page</h1>
        <p>This is my landing page.</p>
        <Link to="/about">About Me</Link>
    </div>
    )
}
```

You can see here that we instead pass `"/about"` to the `to` prop in the `<Link/>` component; this will navigate to our previously created `about` page.

Programmatic navigation

Occasionally, you may need to trigger navigation with something other than a click. Perhaps you need to navigate as a result of a `fetch` request, or when a user submits a form. You can achieve this behavior by making use of the Gatsby `navigate` function:

```
import React from "react"
import {navigate} from "gatsby"

export default function SomePage() => {
    const triggerNavigation = () => {
        navigate('/about')
    }
    return (
        <div>
            <p>Triggering page navigation via onClick.</p>
            <button onClick={()=> triggerNavigation()}>
                About Me
            </button>
        </div>
    )
}
```

Like the `<Link/>` component, the `navigate` function will only work for navigating to internal pages.

We now have a basic Gatsby site set up with the ability to navigate between pages.

Summary

I appreciate that most of the content in this chapter has been theoretical, but it's important to understand the "why" as well as the "how." In this chapter, we have cemented the baseline knowledge of what Gatsby is and grasped the guiding principles we will be using in further chapters to build our website. We've seen examples of where Gatsby is being used and the benefits it can bring. We discussed what dependencies you need and how to initialize Gatsby projects. We have also set up a complete basic Gatsby project and created the first few pages of our website. We then used the built-in Gatsby components and functions to link our pages together.

We will be referencing the theory we've outlined in this chapter throughout this book, but for now, let's turn our focus to styling our web application. In the next chapter, we will identify various different styling methodologies and make an informed choice about which one you should use for your project.

2
Styling Choices and Creating Reusable Layouts

Gatsby sites can be styled in a multitude of ways. In this chapter, we will introduce you to a large selection of styling techniques to help you make an informed choice about how you would like to style your site. Once you've settled on a styling method, we will implement it on the pages we created in *Chapter 1, An Overview of Gatsby.js for the Uninitiated*, before creating the **reusable components** that will be used across all our site pages.

In this chapter, we will cover the following topics:

- Styling in Gatsby
- Styling with CSS
- Styling with Sass
- Styling with Tailwind.css
- Styling with Styled components
- Creating a reusable layout

Technical requirements

In order to navigate this chapter, you will need to have completed the Gatsby setup and created the pages in *Chapter 1, An Overview of Gatsby.js for the Uninitiated.*

In this chapter, we will start adding our first reusable components to our pages. As these components are not standalone pages, we will need a new place to store them. Create a subfolder inside your `src` folder called `components` that we can use.

The code present in this chapter can be found at `https://github.com/ PacktPublishing/Elevating-React-Web-Development-with-Gatsby-4/ tree/main/Chapter02`.

Styling in Gatsby

This chapter is all about styling your Gatsby site, but what does styling refer to? While our React code is defining the structure of our web documents, we will use styling to define our documents' look and feel through page layouts, colors, and fonts. There is an abundance of tools you can use to style any Gatsby project. In this book, I will introduce you to four different approaches – **vanilla CSS, Sass, Tailwind.css**, and **CSS in JS**. Let's explore each of these in a little more detail before deciding which to use.

Vanilla CSS

When your browser navigates to a site, it loads the site's HTML. It converts this HTML into a **Document Object Model** (**DOM**). After this, the browser will begin to fetch resources referenced in the HTML. This includes images, videos, and, more importantly right now, CSS. The browser reads through the CSS and sorts selectors by element, class, and identifiers. It then goes through the DOM and uses the selectors to attach styles to elements as required, creating a render tree. The visual page is then shown on the screen by utilizing this render tree. CSS has withstood the test of time, as we have been shipping CSS in this way with HTML for 25 years. But using vanilla CSS has some pros and cons.

The pros for using vanilla CSS are as follows:

- **Its age**: Because CSS has been around for 25 years at the time of writing this book, there is an abundance of content available on CSS. Due to its age, the chances that someone has already worked out how to fix any issue you encounter is also very high. Both these reasons make vanilla CSS a great choice for a beginner.

- **Understandable syntax**: The syntax that makes up CSS consists of very few abbreviations. Reading it as a beginner, it is far easier to learn what any line of CSS is doing compared to the other style implementations in this chapter.

The cons for using vanilla CSS are as follows:

- **Long style sheets**: In traditional websites, you often see that they only ship with one CSS file. This makes maintaining and organizing styles very difficult, as the file can end up incredibly long. This can lead to a pattern where lazy developers who can't find the styles they are looking for might just append them to the bottom of the file (also known as an **append-only style sheets**). If they do this and the file already exists, then they are just increasing the file size for nothing.

- **Class reuse confusion**: Reusing styles can sometimes lead to more trouble than it's worth. Let's say you have used one specific class in use across various elements in your application. You might update this class to make it fit one instance of it, only to break all the others. If you fall into this cycle multiple times, it can really slow down your development. This can be avoided with a little forward-thinking – instead of reusing classes, make components that are reused. Another option is to create "utility classes" that are unlikely to change; if you'd rather not create these yourself, you should read the section on Tailwind CSS.

- **Inheritance pain points**: By using inheritance, we end up tightly coupling our CSS to the structure of our HTML. If you break that structure, your CSS may no longer work. While inheritance sometimes is unavoidable, we should try and keep it to a minimum.

CSS has withstood the test of time and is still a solid choice today. You might be asking why these are cons when I have listed ways to work around/avoid all of them. These cons can all be fixed one way or another using one of the other implementations in this chapter.

Sass

Sass is a preprocessor scripting language that compiles into CSS. It gives developers tools that allow them to create more efficient CSS.

The pros for using Sass are as follows:

- **Large toolset**: Sass contains a bundle of powerful tools that you can't utilize in vanilla CSS. While we won't be covering these in detail, these include tools like mixins, loops, functions, and imports that can be used to create more powerful and performant CSS for your application. This is a huge pro for Sass.

- **Modules**: You can split your Sass into separate `.scss` files to break down files. You can then import them into one another as needed. This drastically helps improve the organization of your code.

- **Freedom**: Sass enforces a convention of how to write it – you can choose. This means that you can choose a style that suits your team.

The cons for using Sass are as follows:

- **Skip foundations**: Freedom can also be a negative for developers new to styling. If you haven't used Sass before, you may create code that works but in a way that is overly complicated. This can lead to future developers struggling with the code. Concrete CSS guidelines can help avoid this misuse.

- **Naming conventions**: Naming every class you create for every element you style is a tedious process. There are methodologies that can help you create sensible class names; however it still takes a long time.

- **Two sources of truth**: When you write your HTML, you probably will also add class names to your elements to style them. You then jump across to your Sass file to add these class names, only to forget what names you had for them. Jumping back and forth between your HTML and Sass can be an annoying context switch. You might consider abstracting styles away from your markup to be a good thing, but when markup and styles are so interconnected, this can be an inconvenience.

Although Sass is a powerhouse, increased power does also mean increased complexity. While the learning curve may be higher for beginners, gaining control of it will give you a great deal of freedom.

Tailwind (utility-first CSS framework)

Tailwind CSS is a utility-first CSS framework. The "utility-first" approach was created to combat the cons we talked about previously with CSS and Sass. In this methodology, we use small utility classes to build a component's style instead of defining our own class names. It can feel a little like writing inline styling, as your elements will have a string of utility classes added to them, but the benefit is that you don't have to write a single line of your own CSS if you don't want to.

The pros for using Tailwind are as follows:

- **One source of truth**: When using CSS or Sass, you must switch between two files: your markup and your style sheets. Tailwind does away with this concept and instead allows you to embed your styles directly in your markup.

- **Naming conventions**: Tailwind removes the need for you to create your own classes. It has its own classes that are incredibly granular called "utility classes." You use these classes to build up your elements' styles and not worry about creating unique classes for each component.

- **Smaller CSS**: Tailwind provides you with a complete set of utility classes that you rarely need to supplement with your own styles. Your CSS, therefore, stops increasing; in fact, it gets smaller. When you're ready to production-build your application, you can use Tailwind's built-in purge function to remove unused classes.

- **No side effects**: As we are adding styles in our markup and not manipulating the underlying class names, there are never any unintended side effects elsewhere in our application.

The cons for using Tailwind are as follows:

- **Markup legibility**: As your markup contains your style built from utilities, the class names of elements can become incredibly long. When you add in the fact that these may need to change on hover or at breakpoints, your line length can end up very long.

- **Learning curve**: Utility-first requires you to learn many class names to know what tools you must build your styles with. This learning can take some time and slow you down at the beginning, but I believe once you have these under your belt, your development speed will become much faster.

Tailwind hits a great balance of abstraction and flexibility. It is the newest implementation on this list and my personal favorite.

CSS in JS

CSS in JS gives you the ability to write plain CSS within your components while removing the possibility of naming collisions with class names. For the purpose of exploring this option, I will be reviewing the pros and cons of the most popular solution, Styled Components (`https://styled-components.com`). It is, however, worth mentioning that there are many different CSS in JS solutions, including Emotion (`https://emotion.sh`) and JSS (`https://cssinjs.org`).

The pros for using Styled Components are as follows:

- **One source of truth**: Like Tailwind, Styled Components also removes context-switching, as your CSS code is housed within the same file as the component making use of it.

- **Styles tied to components**: Styles are created for use by one specific component and are located next to the markup that implements them. As such, you know exactly what makes use of these styles but, more importantly, you know that editing these styles will only affect the markup located with them.

- **JS in CSS**: We can make use of JS inside our CSS to determine styling. This makes handling conditionals within styles much easier, as we do not have to create two different class names and use a ternary operator.

- **Extending**: It can often be the case that you may want to use a component style but subtly modify it for a different use case. Instead of copying the styles again and creating a new component from the ground up, we can instead create a component that inherits the styling of another.

The cons for using Styled Components are as follows:

- **Performance**: When parsing your styles into plain CSS, Styled Components adds these as style tags in the head of your `index.html`. Styles used across all your pages are pulled in on every page without any way to easily split them. Even caching the styles is difficult, as class names are dynamically generated and can, therefore, change between builds.

If you like a single source of truth, Styled Components improves the legibility of your markup when you are combining everything into one file. While performance is listed as a con, this is something that the community behind Styled Components is making an effort to improve.

Picking a styling tool

When it comes to styling your Gatsby site, there is no right or wrong way of styling it. It will totally depend on your existing skill set, how coupled you want your styles and JS to be, and your own personal preferences. I thought I would end this section by looking at a few common scenarios and what styling implementation I would use for each of them:

- *My experience with styling is limited*: If you are new to styling applications, I would suggest using vanilla CSS. The fundamentals you will learn using this implementation are built on in every other implementation. By learning the basics, you will be able to pick another implementation more easily in the future.

- *I don't want to spend lots of time styling my application*: If you are looking for the option with the least setup, then look no further than Tailwind. Using utility classes will save you a lot of time, as you do not need to create your own classes.

- *I don't like context switching*: In this case, I would lean toward Styled Components or Tailwind, as in both implementations your styles are located next to your markup – one file and one source of truth.

- *I have used CSS and want to build on that*: Using Sass would be a great option for you, as you can write the CSS you know and love but also enhance it with the Sass toolset.

At this point, you should feel ready to make an informed choice about which styling tool is for you. I strongly suggest that you *only implement one of the styling choices outlined in this chapter* instead of trying to mix and match. If you add multiple styling implementations, you can end up in a position where your site styles don't seem to match up. This is because one implementation can override another. By sticking to one method, you have the added benefit of keeping your site's style consistently uniform, which is important, as it reinforces your brand.

Now that you have made a decision, let's start looking at implementations.

Styling with CSS

In this section, we will learn how to implement CSS styling into our Gatsby project.

There are two different methods to adding global CSS styling to our Gatsby site – creating a wrapper component or using gatsby-browser.js.

Creating a wrapper component

The idea behind a wrapper component is to wrap our page components in another component that brings common styles to the page:

1. Create StyleWrapper.css in your components folder:

```
html {
  background-color: #f9fafb;
  font-family: -apple-system, "Segoe UI", Roboto,
    Helvetica, Arial, sans-serif,
    "Apple Color Emoji", "Segoe UI Emoji", "Segoe UI
    Symbol";
}
```

In the preceding code, we are defining a background color and a font family that all children of the HTML tag can inherit.

2. Let's now add some h1 styles to this file:

```
h1 {
  color: #2563eb;
  size: 6rem;
  font-weight: 800;
}
```

Here, we are adding the color, size, and weight of the largest `heading` tag.

3. Similarly, we can also add some styles for the `p` and `a` tags:

```
p {
  color: #333333;
}

a {
  color: #059669;
  text-decoration: underline;
}
```

Here, we are adding a color to each tag and, in the case of the `a` tags, an underline to make them more prominent.

4. Create `StyleWrapper.js` in your `components` folder:

```
import React from "react"
import "./StyleWrapper.css"

const StyleWrapper = ({children}) => (
    <React.Fragment>{children}</ React.Fragment>
)

export default StyleWrapper
```

As the name might suggest, we will use this component to wrap our pages to apply the styles we are importing on the second line.

5. In order to use `StyleWrapper.js`, we need to import it into our pages; let's look at `pages/index.js` as an example:

```
import React from "react"
import {Link} from "gatsby"
import StyleWrapper from "../components/StyleWrapper"

export default function Index(){
    return (
        <StyleWrapper>
            <h1>My Landing Page</h1>
            <p>This is my landing page.</p>
```

```
            <Link to="/about">About me</Link>
        </StyleWrapper>
    )
}
```

In the preceding code, we can see we have imported the styled wrapper on the third line. We then replaced the `div` wrapping with our new layout component. The contained h1, p, and Link elements will be passed into the `StyleWrapper` component as children.

Using gatsby-browser.js

If you want the same styles applied to every page, you might feel that importing `StyleWrapper` on all page instances doesn't feel like you're following **Don't Repeat Yourself** (**DRY**) principles. In cases where you are absolutely sure the styles are needed on every page, we can add them to our application using the Gatsby browser instead:

1. Create a `styles` folder inside your `src` directory. As these styles are being used globally and are not tied to a specific component, it does not make sense to store them in the `component` directory, as we did when implementing `StyleWrapper.js`.

2. Create a `global.css` file in your `styles` folder and add the following:

```css
html {
    background-color: #f9fafb;
    font-family: -apple-system, "Segoe UI", Roboto,
        Helvetica, Arial, sans-serif,
        "Apple Color Emoji", "Segoe UI Emoji", "Segoe UI
        Symbol";
}
h1 {
    color: #2563eb;
    size: 6rem;
    font-weight: 800;
}
p {
    color: #333333;
}
```

```
a {
    color: #059669;
    text-decoration: underline;
}
```

Here, we are adding the exact same styles that we had in the alternate CSS implementation, so I won't explain them again here. The key difference is in this next step.

3. Navigate to `gatsby-browser.js` and add the following:

```
import "./src/styles/global.css"
```

By importing our CSS in `gatsby-browser.js`, Gatsby will wrap every page with this CSS.

Verifying our implementation

Regardless of which of the two methods you opted for, if everything has gone according to plan, you should be presented with a styled site that looks like this:

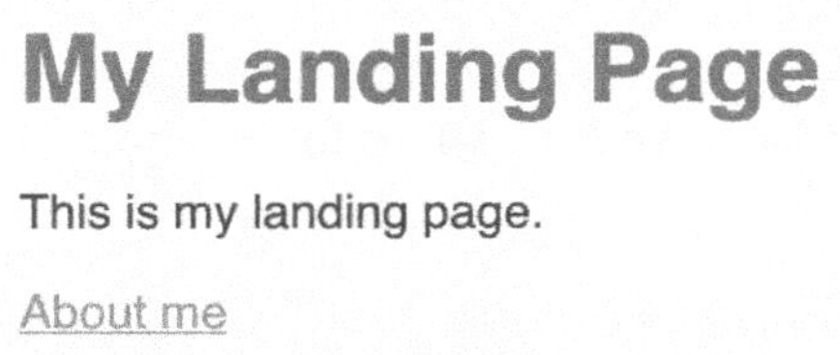

My Landing Page

This is my landing page.

About me

Figure 2.1 – Development of the index page with styles

You should be able to pick out your CSS additions on this page.

You have now implemented CSS as a styling tool within your Gatsby site. You can disregard the other styling implementations that follow and proceed to the *Creating a reusable layout* section.

Styling with Sass

In this section, we will learn how to implement Sass styling in our Gatsby project:

1. To start using Sass, we will need to install it along with a few other dependencies. Open a terminal at the `root` folder of your project and run the following:

```
npm install sass gatsby-plugin-sass
```

Here, we are installing the core Sass dependency as well as the Gatsby plugin that integrates it.

2. Modify your `gatsby-config.js` file with the following:

```
module.exports = {
  plugins: [
    'gatsby-plugin-sass'
  ],
};
```

Here, we are updating our Gatsby configuration to let Gatsby know to make use of the `gatsby-plugin-sass` plugin. Now, create a `styles` folder inside your `src` directory.

3. Create a `global.scss` file in your `styles` folder and add the following:

```
html {
  background-color: #f9fafb;
  font-family: -apple-system, "Segoe UI", Roboto,
    Helvetica, Arial, sans-serif,
    "Apple Color Emoji", "Segoe UI Emoji", "Segoe UI
    Symbol";
}
```

I rarely add more than HTML styles to the `global.scss` file. Instead, I prefer to import other `.scss` files into this one. This keeps styles in order and the files small and readable. As an example, let's create `typography.scss` to store some typography styles:

```
h1 {
  color: #2563eb;
  size: 6rem;
  font-weight: 800;
}
p {
  color: #333333;
}

a {
  color: #059669;
```

```
      text-decoration: underline;
   }
```

4. Here, we are adding a color to each and, in the case of the a tags, adding an underline to make them more prominent. We can now import this file into our `global.scss` file:

```
@import './typography';
html {
  background-color: #f9fafb;
  font-family: -apple-system, "Segoe UI", Roboto,
    Helvetica, Arial, sans-serif,
    "Apple Color Emoji", "Segoe UI Emoji", "Segoe UI
      Symbol";
```

5. Navigate to your `gatsby-browser.js` file and add the following:

```
import "./src/styles/global.scss";
```

This tells our Gatsby application to include this style sheet on the client, allowing us to make use of it in our application.

You have now implemented Sass as a styling tool within your Gatsby site. You can disregard the other styling implementations that follow and proceed to the *Creating a reusable layout* section.

Styling with Tailwind.css

In this section, we will learn how to implement Tailwind styling in our Gatsby project:

1. To start using Tailwind, we will need to install it along with a few other dependencies. Open a terminal at the `root` folder of your project and run the following:

```
npm install postcss gatsby-plugin-postcss tailwindcss
```

Here, we are installing PostCSS, its associated Gatsby plugin, and `tailwindcss`. PostCSS is a tool for transforming styles with JS plugins. These plugins can lint your CSS, support variables and mixins, transpile future CSS syntax, inline images, and more. In the case of Tailwind, there is a specific tailwind plugin for PostCSS that we will be implementing.

2. Modify your `gatsby-config.js` with the following:

```
module.exports = {
  plugins: [
    'gatsby-plugin-postcss'
  ],
};
```

Here, we are updating our Gatsby configuration to let it know to make use of the Gatsby PostCSS plugin.

3. In order to use PostCSS, it requires `postcss.config.js` to be present at the root of your project. Go ahead and create this file now and add the following:

```
module.exports = () => ({
  plugins: [require("tailwindcss")],
});
```

In this file, we are telling PostCSS to make use of our newly installed `tailwindcss` package.

4. Much like PostCSS, Tailwind also requires a configuration file. Tailwind has a built-in script for creating the default configuration. Open a terminal and run the following:

```
npx tailwindcss init
```

If this command is successful, you should notice that a new `tailwind.config.js` has been created at the root of your project. The default configuration within this file will work just fine, so for now, we don't need to edit it.

5. Create a `styles` folder inside your `src` directory.

6. Create a `global.css` file inside your `styles` folder and add the following:

```
@tailwind base;
@tailwind components;
@tailwind utilities;
```

7. Add the following to the `gatsby-browser.js` file:

```
import "./src/styles/global.css";
```

This tells our Gatsby application to include this style sheet on the client, allowing us to make use of Tailwind classes.

With these steps concluded, we now have everything in place to start using Tailwind within our application. To make use of Tailwind's utility classes, we can use React's `className` prop within our components; for example, in `pages/index.js`, we could add the following:

```
import React from "react"
import {Link} from "gatsby"

export default function Index(){
    return (
        <div>
            <h1 className="text-3xl font-bold text-blue-
                600">My Landing Page</h1>
            <p>This is my landing page.</p>
            <Link to="/about">About me</Link>
        </div>
    )
}
```

In the preceding code, we are modifying the style of the heading with the following utility classes:

- `text-3xl`: Set the text to the third extra-large size, equivalent to 1.875 rem.
- `font-bold`: Set the text to the bold font-weight.
- `text-blue-600`: Set the color of the text to blue.

You can alternatively append styles to the `global.css` file that we created to have them included:

```
@tailwind base;
@tailwind components;
@tailwind utilities;

h1 {
    @apply text-3xl font-bold text-blue-600;
}
```

Here, you will see the exact same styles, just defined globally. Both will equate to the same styling on the h1 tag; deciding which variation to use is all about frequency. If you intend to use this h1 style more than once, you should be incorporating it into your CSS to save you writing the same styles over and over.

Let's now supplement this with a few more styles:

```
@tailwind base;
@tailwind components;
@tailwind utilities;

h1 {
    @apply text-3xl font-bold text-blue-600;
}

p {
    @apply text-gray-800;
}

a {
    @apply text-green-600 underline;
}
```

Here, we are adding a color to each element and, in the case of the a tags, adding an underline to make them more prominent.

You have now implemented Tailwind as a styling tool within your Gatsby site. You can disregard the other styling implementations that follow and proceed to the *Creating a reusable layout* section.

Styling with styled-components

In this section, we will learn how to implement Styled Components as a styling tool in our Gatsby project:

1. Open a terminal at the root folder of your project and run to install your dependencies:

    ```
    npm install gatsby-plugin-styled-components styled-
    components babel-plugin-styled-components
    ```

These are the details of the dependencies:

a.   `styled-components`: The Styled Components library

b.   `gatsby-plugin-styled-components`: The official Gatsby plugin for Styled Components

c.   `babel-plugin-styled-components`: Provides consistently hashed class names between builds

2.   Update your `gatsby-config.js` with the following:

```
module.exports = {
  plugins: ['gatsby-plugin-styled-components'],
}
```

This instructs Gatsby to use the Styled Components plugin that we just installed.

We can have all the pieces in place to create styles on a page/component level and a global level.

3.   To demonstrate utilizing them, navigate to your `pages/index.js` file and add the following:

```
import React from "react"
import {Link} from "gatsby"
import styled from "styled-components";

const Box = styled.div'
  background-color: #333;
  padding: 20px;

  h1 {
    color: #fff;
    margin: 0 0 10px;
    padding: 0;
  }

  p {
    color: #fff
  }
'
```

```
export default function Index(){
    return (
        <Box>
            <h1>My Landing Page</h1>
            <p>This is my landing page.</p>
            <Link to="/about">About me</Link>
        </Box>
    )
}
```

Here, we have defined a component that applies styling to a `div` tag. We can see that it also has styles for any `h1` or `p` tag that are children.

4. Occasionally, you may want to create styles globally; in order to demonstrate this, navigate to your `gatsby-browser.js` file and add the following:

```
import React from "react"
import { createGlobalStyle } from "styled-components"

const GlobalStyle = createGlobalStyle'
  body {
    background-color: ${props => (props.theme ===
    "blue" ? "blue" : "white")};
  }
'

export const wrapPageElement = ({ element }) => (
    <>
        <GlobalStyle theme="blue"/>
        {element}
    </>
)
```

We use the `styled-components` `createGlobalStyle` helper function to create our global styles. This stops Styled Components from being scoped to a local CSS class.

By using the `wrapPageElement` method, we tell Gatsby to wrap every page in the component. We can make use of this to wrap every page in our global styles.

Regardless of your implementation choice, you should now have the basics in place to start building a fully styled site. Let's now start creating the reusable layout that we will utilize across the site.

Creating a reusable layout

Most websites feature headers and footers that are present across all their pages. With our knowledge of how pages work, you might be tempted to import a header component into every page component. But wait – what happens when you suddenly need to pass that component a new prop? Situations like these are why it's a good idea to reduce any duplication across pages. Instead, it's a much better option to create a layout component that contains a header and footer that we can then wrap our pages in.

In order to keep our `components` folder well structured, it's useful to create subfolders to house different parts of the site. Create a `layout` folder in the `components` folder to house components that are related to the layout. We will use these layout components across all our page files. Now, let's populate this folder with a header, footer, and layout component.

> **Important Note**
> In the code examples in this section, you will notice I am using `Tailwind.css` to style my components. In the accompanying GitHub repository (`https://github.com/PacktPublishing/Elevating-React-Web-Development-with-Gatsby-4/tree/main`), you can find implementations of these components using all the styling implementations we have covered in this chapter. In future chapters, I will be sticking to Tailwind.

Site header

A header component acts as the anchor of our site. It is common to include your site header across all your pages so that visitors are reminded that they are on your site.

To get started, let's create a `Header.js` component in our `components` folder:

```
import React from "react"

const Header = () => (
    <header>
        <p>Site Header</p>
    </header>
```

```
)

export default Header
```

In the preceding code, we are creating the most basic of header examples. Note that we are making use of the HTML header tag. As we will learn in *Chapter 6, Improving Your Site's Search Engine Optimisation*, using the correct tags when creating content is very important, as it helps web crawlers and accessibility tools understand your site.

Site footer

Adding a footer to your site can be a powerful tool. I like to think of it as a way to keep user engagement after they have finished a page. We can use it to give people quick links to our social media so that they can get in touch, we can suggest other interesting content that they might enjoy, and we can even tell them how many views the current page has had.

Let's get started with a basic implementation. Create a Footer.js component in the components folder:

```
import React from "react"

const Footer = () => (
    <footer>
        <p>Site Footer</p>
    </footer>
)

export default Footer
```

Much like our Header, it's important that we use the proper HTML footer tag.

Layout component

We could directly import our header and footer onto every page we create, but if we did that, it would lead to a lot of duplication. One common way to get around this is to create a `Layout` component. We wrap every page we build in this component. Not only is this an easy way to bring in our header and footer but it also allows us to style the main content of every page with minimal effort:

```
import React from "react"
import Footer from "./Footer"
import Header from "./Header"

const Layout = ({children}) => (
    <div>
        <Header/>
        <main>
            {children}
        </main>
        <Footer/>
    </div>
)

export default Layout
```

Here, you can see I am importing our newly created `Header` and `Footer` components. I am making use of the `children` prop and rendering child content within a `main` block.

To demonstrate using the `Layout` component, let's wrap our index page in the component. Modify your `index.js` in pages with the following:

```
import React from "react"
import {Link} from "gatsby"
import Layout from "../components/layout/Layout"

export default function Index(){
    return (
        <Layout>
            <h1>My Landing Page</h1>
            <p>This is my landing page.</p>
            <Link to="/about">About me</Link>
```

```
        </Layout>
    )
}
```

You can see that I have wrapped the landing page in our new `Layout` component that has been imported on the third line. If you spin up `gatsby develop` at this point, you should see your page content with a header before it and a footer after it. You can continue to wrap your other pages in your `layout` component at this moment. Before proceeding, let's step back for a moment and look at how we can organize the components we're creating for our pages.

> **Tip**
> Some of the styling implementations discussed previously make use of style wrappers. If your implementation has made use of a style wrapper, import it into your `layout` component and wrap content in this component. This way, you only wrap your pages in one component instead of both the `layout` and `style wrapper` components.

Organization with atomic design

As your site expands, it's important to try and keep your `components` folder structured. One commonly used method is to use **atomic design** principles. Atomic design is the process of creating effective interface design systems by breaking down your site elements into atoms, molecules, organisms, templates, and pages:

- **Atoms**: These are the smallest components that might be incorporated into our site, such as buttons, typography components, or text inputs. Atoms cannot be logically broken down into subcomponents while still keeping them functional.

- **Molecules**: Built from two or more atoms, molecules are small groups of elements that work together to provide some functionality. A search box consisting of a text input and a button could be considered a molecule.

- **Organisms**: Built from a group of molecules and atoms, these form larger sections of an interface, such as a site's hero section.

- **Templates**: These wrap organisms in a layout and provide the content structure and skeletal structure of a page.

- **Pages**: An instance of a template with real-world content in place.

Using atomic design when building components allows you to break your components down into smaller self-contained units. These can be tested and developed in isolation before importing them into your application, allowing for a more rigorous development process but also reducing the reliance on backend logic when conducting frontend development.

Once we have defined our atomic design pattern, we can be far more flexible when working with styling. Changing an atom's style will also update the styles that any molecules and organisms were making use of.

Depending on your styling implementation, it can also be a good idea to abstract commonly used tokens, such as brand colors, spacing rules, and font families. It's much easier to maintain a project with a single source of truth, rather than darting all over your application pasting hex values to modify your brand colors.

Using atomic design to organize your `components` folder can really help as things scale, so keep it in mind in future chapters as your application expands.

Summary

In this chapter, you have learned how you can style a Gatsby site in a multitude of ways. This should aid you in making an informed choice about what way you will style your application going forward. We have seen how to style your Gatsby site using CSS, Sass, Tailwind.css, and Styled Components. You should have decided on one of these and implemented them. In future chapters, I will use Tailwind.css to style the application, but this is a personal preference. You should use whatever you feel is best for your site and your existing knowledge.

We also started creating the first reusable components that will make up the backbone of our site. While our `layout` component may seem primitive right now, we will be integrating it with content in the next chapter and adding imagery to bring it to life even further in *Chapter 5, Working with Images.*

Before continuing to the next chapter, I encourage you to spend time building on the styles outlined here until you have your existing pages looking how you want them to. While I think it's best to define your own styles, you can find an example of a completely styled version of the site in Tailwind.css in the code repository.

In the next chapter, we will begin sourcing content from local files, CMSes, and APIs. We will be using this data to start programmatically creating pages on our Gatsby site.

3
Sourcing and Querying Data (from Anywhere!)

In this chapter, you will learn about Gatsby's data layer. You will start by understanding what we mean by data in the context of Gatsby before learning the basics of **GraphQL**. Once you have this understanding, you will learn how to source and query data from local files. We will then look at sourcing data from a couple of Headless CMSes.

In this chapter, we will cover the following topics:

- Data in Gatsby
- Introducing GraphQL
- Sourcing and querying data from local files
- Sourcing and querying data from a Headless CMS

Technical requirements

To complete this chapter, you will need to have completed *Chapter 2*, *Styling Choices and Creating Reusable Layouts*.

The code for this chapter can be found at `https://github.com/PacktPublishing/Elevating-React-Web-Development-with-Gatsby-4/tree/main/Chapter03`.

Data in Gatsby

Before diving in, I think it's important to establish what we mean by **data** in the context of this book. When referring to data, we are referring to any medium of static content that is not React code. Up until now, we have been adding text within our React components directly. As a developer, this can be a perfectly acceptable way to build a small site but as things scale up, having content mixed into your markup can make it much harder to develop. It also makes it impossible for colleagues without React experience to update or add new content to the site.

It is a much more common practice to store data that's separate from our pages and components, pulling it in as required. There are two ways in which we can store this data:

- **Locally**: Files stored alongside our source code in the respective repository, such as JSON, CSV, Markdown, or MDX files.

- **Remotely**: Files stored in another location that we ingest as part of our build processes, such as content from a Headless CMS, database, or API.

> **Important Note**
>
> You may have noticed the absence of images being referenced when talking about data and might be wondering how to work with them. Due to their complexity, images have a dedicated chapter in this book – *Chapter 5*, *Working with Images*.

Now that we understand what we mean by data in Gatsby, let's learn how we can query it within our application so that we can use it on our site pages.

Introducing GraphQL

GraphQL is a specification for querying data – general guidelines on how to query data efficiently. This specification was developed by engineers at Facebook in 2012 while working on their mobile application's **REST** services. They wanted to use their existing REST service on their mobile platforms, but it was going to require heavy modification and specific logic for mobile platforms in various areas of their APIs. The engineers also noticed that there were lots of data points in the responses to their API requests that they were not using. This meant that people with low network bandwidth were loading data they weren't even using.

So, the team at Facebook started to work on GraphQL to solve these problems and rethink the way they could fetch data for devices. GraphQL shifted the focus from the backend engineers specifying what data is returned by what request, to the frontend developers specifying what they need.

GraphQL for Gatsby

Gatsby always uses GraphQL whenever you want to get data from within it. This is a great feature as we have an efficient way of getting data, regardless of its type. Gatsby can call GraphQL APIs directly if you already have a GraphQL server set up. However, a lot of the data we need to use on the web is not already in GraphQL format.

Luckily, Gatsby's plugin architecture allows you to get non-GraphQL data into Gatsby, then use GraphQL to query it once you have it there. Regardless of whether your data is local or remote, or what format it is in, you can use one of Gatsby's plugins to pull the data. Then, you can use the GraphQL specification to query for that data on our pages.

This is a great architecture that works with all of our content, no matter where it comes from. When it gets into Gatsby, we always query and retrieve that data in the same way.

Let's look at a high-level example of what a GraphQL query contains:

```
query SampleQuery {
  content {
    edges {
      node {
        property
      }
    }
  }
}
```

Here, you can see that we used the word `query`, followed by the name of the query, which, in our case, is `SampleQuery`. Then, inside of the curly braces, we specify what kind of content we want to get – where you see `content` here, this would change to be the source of content you want. `edges` refers to a collection of connected items within that content source that have a relationship returned as an array. Then, when we go a level deeper, we have `node`, which refers to an individual item. Here, you can see that we are querying a single property.

One of the great things about GraphQL is that you can be very specific about the data you need and only get that specific content. As shown in the previous example, we are only querying a single property of the node, but what if it contained a hundred properties instead? By pulling out only exactly what we need, we can create a very specific query that only gets us what we need.

Now, let's look at a Gatsby-specific GraphQL query:

```
query MySitePages {
  allSitePage {
    edges {
      node {
        path
      }
    }
  }
}
```

Here, we can see that we are naming the query `MySitePages`. The content we are retrieving is from the `allSitePage` source, which is a default collection that contains all the pages that have been created within a Gatsby project. `edges` refers to all the pages, whereas `node` refers to a specific page we want. Inside each page, we are querying for the `path` parameter of that page.

When running this query in Gatsby, it is going to return JSON. If you run the preceding query within our site and log the result, you would see the following object:

```
{
  "data": {
    "allSitePage": {
      "edges": [
        {
          "node": {
```

```
          "path": "/404/"
        }
      },
      {
        "node": {
          "path": "/about/"
        }
      },
      {
        "node": {
          "path": "/"
        }
      }
    ]
  }
 }
}
```

As you can see, what we get back is an object with a data property. Within that, you can see our named query and its edges. The edges contain each node and its corresponding path property. Within the result, we can see every page that exists on the site – we have our 404 page, the about page, and the home page.

Now, let's learn about filtering and sorting data within GraphQL.

Filtering in GraphQL

Sometimes, all the nodes of the returned data are not useful. We may occasionally want to filter out nodes based on a particular field. Let's take a look at an example where we are filtering out nodes from our allSitePage source:

```
query AllSitePagesExcept404 {
  allSitePage(filter: {path: {ne: "/404/"}}, limit: 1) {
    edges {
      node {
        path
      }
    }
```

```
    }
  }
```

Here is an example in which we get a single page where the path does not equal (ne for short) /404/. Filtering is something we will look at in more detail as we start to develop more complex queries for our pages. Right now, it's important just to recognize that it is possible.

In Gatsby, it is possible to obtain a single node on its own, but it is more common to query a collection. For example, if we wanted to retrieve a single SitePage node, we could use the following query:

```
query ASingleSitePage {
  sitePage {
    path
  }
}
```

This query will receive the first node that matches the request and return it as an object instead of a larger array.

Now that we understand how GraphQL queries are constructed, let's take a look at how we can use GraphiQL to explore our data.

Using GraphiQL

When it comes to learning GraphQL, it's fortunate that Gatsby ships with a tool called GraphiQL (https://github.com/graphql/graphiql). This is a web interface that hooks up to all of the GraphQL options in Gatsby and gives us a nice interface for testing and playing around with queries before we embed them into our code.

As we know, when developing our site, Gatsby opens up http://localhost:8000 to preview our site while we are building it. If you navigate to http://localhost:8000/___graphql, you will pull up a GraphiQL interface that is connected to your development Gatsby site. When you open this page, you should be presented with something that looks like this:

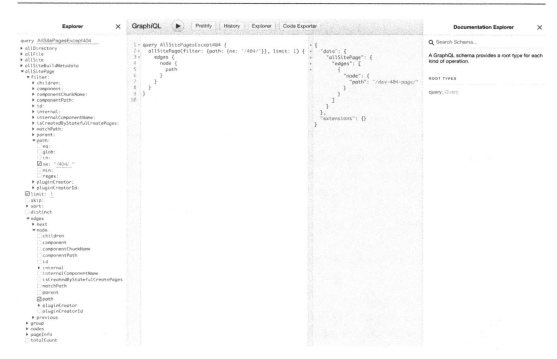

Figure 3.1 – GraphiQL user interface

On the far left, you will see **Explorer**, which shows all the possible pieces of content we could get using GraphQL inside of Gatsby. You can check the properties within the **Explorer** area to have GraphiQL automatically build the query for you. In the central left column, we can see the query that we need to use to retrieve the data we want. When you hit the **Play** button above the query, you will see the result of that query on the central right column, with a JSON object containing the data property and our query's result inside it. On the far right, you will see the **Documentation Explorer** area, which you can use as an alternative way to explore your data and identify the different types of data you have available.

Now, let's learn where we can use queries to retrieve data within our application.

Using constructed GraphQL queries

There are three main places where you can use a GraphQL query in your Gatsby projects:

- `Gatsby-node.js`: This file is one of the places we can create pages programmatically based on dynamic data. If we had a list of blog posts in Markdown and we wanted to create a page for each one, we would use a query here to retrieve the data from the posts that we need to dynamically create the pages for.

- **Within pages**: We can append queries to single instance pages to make data available within that page. This is how we will be testing the data we source within this chapter. We can also query inside **page templates**, something we haven't discussed yet, but it is a key concept we will look at in detail in *Chapter 4, Creating Reusable Templates*. A page template could take a slug based on the URL and then run a query based on that URL to work out what page to display. In both single-instance pages and templates, the query is run at build time, so the pages that are created are still static.

- **Within any other component**: We can also retrieve GraphQL data within any React component we have created. There is a different method to retrieving data outside of page templates because outside of a page template, you cannot get dynamic content using variables. As such, queries run this way are static. We will see examples of static queries in *Chapter 5, Working with Images*.

Now that you understand the basics of GraphQL in Gatsby, let's start ingesting different kinds of data into our GraphQL layer.

Sourcing data from local files

In this section, we will learn how to source and query data from local files. As we mentioned previously, when we say local files, we are referring to files located alongside the code in our repository.

Site metadata

A great place to store small reusable pieces of data is within the `gatsby-config.js` file. Gatsby exposes the `siteMetadata` property to the data layer so that you can retrieve it throughout your application. In the context of our website, I would suggest storing your website address, your name, your role, and a short bio here. If this is implemented consistently, whenever any of these pieces of information change, you can change the field once in `siteMetadata` and see the change reflected across your whole site.

> **Tip**
> `gatsby-config.js` is a file that you will often find growing quite large as you expand your Gatsby projects. To try and keep things ordered, try and reserve your `siteMetadata` for a handful of small strings. If you are considering adding a large block of text here, it might be better to add it as a Markdown file.

Let's create some site metadata and ingest it on our home page:

1. First, update `gatsby-config.js` with the following code:

```
module.exports = {
  siteMetadata: {
    siteUrl: 'https://your.website',
    name: 'Your Name',
    role: 'Developer at Company',
    bio: 'My short bio that I will use to introduce
         myself.'
  },
  plugins: [
    // your plugins
  ],
};
```

The `siteMetadata` key sits next to the plugins we have defined. Here, you can see we have defined the key values I suggested earlier. Keep in mind that these key values are just a suggestion and that if you want to add or remove keys, feel free to do so.

2. Use the GraphiQL interface to construct the GraphQL query to retrieve the data. This should look like this:

```
query BasicInfo {
  site {
    siteMetadata {
      name
      role
    }
  }
}
```

Your site metadata is available within the `site` source. In the preceding query, we are only retrieving `name` and `role`.

3. Embed your constructed query on your index page:

```
import React from "react";
import { Link, graphql } from "gatsby";
import Layout from "../components/layout/Layout";


export default function Index({ data }) {
  const {
    site: {
      siteMetadata: { name, role },
    },
  } = data;


  return (
    <Layout>
      <div className="max-w-5xl mx-auto py-16 lg:py-
        24">
        <h1 className="text-4xl md:text-6xl font-bold
         text-black pb-4">
          {name}
        </h1>
        <p className="mb-4">{role}</p>
        <Link to="/about" className="btn">
          About me
        </Link>
      </div>
    </Layout>
  );
}


export const query = graphql'
  {
    site {
      siteMetadata {
        name
```

```
        role
      }
    }
  }
';
```

Here, you can see we are importing `graphql` from Gatsby. We are then appending our query from *Step 2* to the end of the file, below our page component. The export name isn't important as Gatsby looks for any GraphQL string within your pages, but here, you can see I am calling it `query`.

When Gatsby builds this page, this query is pulled out of our source code, parsed, and run, and the resultant data is passed into our page component via the data prop you can see on line 5. We can then use the data contained within the query (in our case, `name` and `role` from `siteMetadata`) to populate our site hero.

> **Important Note**
>
> You can only export one query per component. If you ever need more data on the page, instead of exporting another query, extend your existing query.

Now, let's learn about how we can ingest data from sources that are not included with Gatsby out of the box – starting with Markdown.

Markdown

The Markdown syntax is a popular way to write content on a Gatsby site. If you have used GitHub or Bitbucket before, chances are you've already encountered this format as they both make use of it in README files. Markdown is a great format for longer pieces of writing within your site – documentation, blog posts, or even a long bio.

To start using Markdown in Gatsby, you only need to create text files – no additional infrastructure is required to implement it. Gatsby also provides **core-plugin** (a plugin owned and maintained by the Gatsby team) to process Markdown into content that can be used by our components. Using core-plugin, no code is required to implement Markdown and get set up.

Let's create a short biography in Markdown and add it to our about page:

1. Create a folder to store our Markdown called MD at the root of your project.

    It's good practice to keep this folder outside of your src directory as it does not contain any source code but instead is text content. This makes it much easier for a developer without React experience to modify site content.

2. Create a folder inside /MD to store your bio called bio. As we add more Markdown files that serve up different types of content, it's helpful to keep them separated.

3. Create a bio.md file inside our newly created bio folder and add the following code:

    ```
    ---
    type: bio
    ---
    ```

    This is the first part of the file and contains **YAML** frontmatter. YAML is a human-readable data-serialization language. Here, we are defining a type. This type will help us query for this exact file in our GraphQL query.

4. Create the body of your biography using Markdown syntax:

    ```
    ---
    type: bio
    ---

    # A short biography about me
    This is a very short biography about ***me***. But it
    could be as long as I want it to be.
    ```

    You can use any valid Markdown syntax here; I have kept this example brief by just including one heading and a paragraph, but feel free to add as much as you like.

5. Install gatsby-source-filesystem:

    ```
    npm install gatsby-source-filesystem
    ```

    As its name might suggest, this plugin allows Gatsby to read local files.

6. Install gatsby-transformer-remark:

    ```
    npm install gatsby-transformer-remark
    ```

We can use this plugin to recognize Markdown files and read their content. This plugin will read in the syntax and convert it into HTML that we can then embed in our components.

7. Next, let's configure our new dependencies in `gatsby-config.js`:

```
module.exports = {
  siteMetadata: {
    siteUrl: 'https://your.website',
    name: 'Your Name',
    role: 'Developer at Company',
    bio: 'My short bio that I will use to introduce
          myself.',
  },
  plugins: [
    {
      resolve: 'gatsby-source-filesystem',
      options: {
        name: 'markdown-bio',
        path: '${__dirname}/MD ',
      },
    },
    'gatsby-transformer-remark',
    'gatsby-plugin-postcss',
  ],
};
```

Here, we are introducing Gatsby to our new plugins. We are using `gatsby-source-filesystem` to tell Gatsby to read files from the Markdown folder we created previously.

We also added `gatsby-transformer-remark` so that Gatsby can read Markdown files into its GraphQL layer.

8. Start your development server and navigate to your GraphiQL interface. Construct and run the query to retrieve just the `bio` information:

```
query Biography {
  markdownRemark(frontmatter: {type: {eq: "bio"}}) {
    html
  }
}
```

Here, we have constructed a query where we are retrieving the HTML from `markdownRemark`. We are filtering the Markdown where the frontmatter type is equal to `bio` and since there is only one such file, we will always retrieve the correct file. By running this query in the GraphiQL interface, you should see something that looks like this:

```
{
  "data": {
    "markdownRemark": {
      "html": "<h1>A short biography about
        me</h1>\n<p>This is a very short biography
        about <em><strong>me</strong></em>. But it
        could be as long as I want it to be.</p>"
    }
  },
  "extensions": {}
}
```

Here, you can see that the Markdown we wrote in the file has been transformed into HTML that we can now use within our pages.

9. Embed this query in your `about` page:

```
import React from "react";
import { graphql } from "gatsby";
import Layout from "../components/layout/Layout";

export default function About({ data }) {
  const {
    markdownRemark: { html },
  } = data;
  return (
```

```
      <Layout>
        <div className="max-w-5xl mx-auto py-16 lg:py-24
          text-center">
          <div dangerouslySetInnerHTML={{ __html: html
            }}></div>
        </div>
      </Layout>
    );
}


export const query = graphql'
  {
    markdownRemark(frontmatter: { type: { eq: "bio" }
  }) {
      html
    }
  }
';
```

Here, we have appended our query to the bottom of the page and retrieved the resultant data via the data prop. I'd like to draw your attention to the div with the dangerouslySetInnerHTML prop. dangerouslySetInnerHTML is React's replacement for using innerHTML in the browser's DOM.

It's considered *dangerous* because if the content can be edited by a user, this can expose users to a **cross-site scripting attack**. A cross-site scripting attack injects malicious code into a vulnerable web application. In our case, however, the content is always static and always defined by us, so we have nothing to worry about.

Markdown can be a great option if you want to write long-form articles, but what if you want to make your articles more interactive? Maybe you want a poll in the middle of your post or a prompt for users to sign up to your email between two paragraphs? There are plenty of scenarios like these that simply cannot be achieved elegantly in Markdown. For functionalities such as these, MDX is the answer.

MDX

MDX is a format that allows you to enhance your Markdown with JSX. You can import components into your Markdown and embed them in your content.

Let's make an enhanced biography on our about page using MDX that contains your employment history:

1. Create a folder to store our Markdown called MDX at the root of your project. As with Markdown (and for the same reasons), it's good practice to keep this folder outside of src, even though it can contain React components.

2. Create a folder inside /MDX to store your bio called bio (as we did with our Markdown).

3. Create a folder called components within your /MDX folder to store React components specifically for use within our MDX files.

4. Create an EmploymentHistory component in the components folder that we can embed in our mdx file:

```jsx
import React from "react";
const employment = [
  {
    company: "Company One",
    role: "UX Engineer",
  },
  {
    company: "Company Two",
    role: "Gatsby Developer",
  },
];

const EmploymentHistory = () => (
  <div className="text-left max-w-xl mx-auto">
    <div className="grid grid-cols-2 gap-2 mt-5">
      {employment.map(({ role, company }) => (
        <>
          <div className="flex justify-end font-
            bold"><p>{role}</p></div>
          <p>{company}</p>
        </>
```

```
    ))}
   </div>
  </div>
);
```

```
export default EmploymentHistory;
```

I am using employment history as an example here, but this can be any valid React component. In this example, we have defined a small array of employment experiences containing objects, each with a company and role. In EmploymentHistory, we map over those roles and lay them out in a grid. We then export the component as normal.

5. Create bio.mdx in /MDX/bio:

```
---
type: bio
---
import EmploymentHistory from
  "../components/EmploymentHistory";

# A short biography about me
This is a very short biography about **_me_**. But it
could be as long as I want it to be.


### ***My Employment History***


<EmploymentHistory />
```

Like our Markdown, we can include frontmatter in MDX files. Here, we are once again specifying type as bio. Just below that, you will see we have introduced an import statement pointing to our newly created component. We can then use the imported component wherever we like within the body of our content, much like I have on the last line in the preceding example.

6. Install the necessary mdx dependencies:

```
npm install gatsby-plugin-mdx @mdx-js/mdx @mdx-
js/react
```

7. Configure `gatsby-config.js` so that it includes the `gatsby-plugin-mdx` plugin:

```
module.exports = {
  siteMetadata: {
    siteUrl: 'https://your.website',
    name: 'Your Name',
    role: 'Developer at Company',
    bio: 'My short bio that I will use to introduce
          myself.',
  },
  plugins: [
    {
      resolve: 'gatsby-source-filesystem',
      options: {
        name: 'mdx-bio',
        path: '${__dirname}/MDX ',
      },
    },
    'gatsby-plugin-mdx',
    'gatsby-plugin-postcss',
  ],
};
```

We use `gatsby-source-filesystem` to tell Gatsby to read files from the MDX folder we created previously. We have also added `gatsby-plugin-mdx` so that Gatsby can read MDX files into its GraphQL layer.

8. Start your development server and navigate to your GraphiQL interface. Construct and run the query to retrieve our updated MDX bio:

```
query Biography {
  mdx(frontmatter: { type: { eq: "bio" } }) {
    body
  }
}
```

Here, we have constructed a query where we are retrieving the `mdx` body from the `mdx` source, where the frontmatter type is equal to `bio`.

9. Embed the query in your about page:

```
import React from "react";
import { graphql } from "gatsby";
import Layout from "../components/layout/Layout";
import { MDXRenderer } from "gatsby-plugin-mdx";

export default function About({ data }) {
  const {
    mdx: { body },
  } = data;
  return (
    <Layout>
      <div className="max-w-5xl mx-auto py-16 lg:py-24
        text-center">
        <MDXRenderer>{body}</MDXRenderer>
      </div>
    </Layout>
  );
}

export const query = graphql`
  {
    mdx(frontmatter: { type: { eq: "bio" } }) {
      body
    }
  }
`;
```

Here, we have appended our query to the bottom of the page and retrieved the resultant data via the data prop. We then used MDXRenderer from gatsby-plugin-mdx to render the MDX body's content.

> **Important Note**
> Using MDXRenderer does increase your bundle size and the time it takes for your JavaScript to be parsed. This is because instead of rendering all the HTML at build time, any pages containing MDX are now being rendered to HTML on the frontend. This is important to keep in mind as it will negatively impact your site's performance.

Now that we understand how to ingest local data, let's look at sourcing data from a remote source – a **Content Management System (CMS)**!

Sourcing data from a Headless CMS

A Headless CMS is a CMS that purely focuses on the content itself and does not care about how it's presented. Traditional CMSes store content in a database and then use a series of HTML templates to control how content gets presented to viewers. In Headless CMSes, however, instead of returning HTML, we return structured data via an API.

Content creators can still add and edit data via a user interface, but the frontend is stored completely separately. This is perfect for when your content creators are not developers, or when you're out and about and want to write a post on your phone without having to spin up your laptop.

With Gatsby's vast plugin ecosystem, your site can support many different Headless CMSes with very little effort. You could write a book on how to implement every one of them into your project, so, instead, let's focus on two – GraphCMS and Prismic.

> **Important Note**
> *Only implement one of the Headless CMS choices outlined in this chapter.* Not only would having two different sources for the same type of data be confusing, but it would also lead to longer site build times as data will need to be retrieved from two sources instead of one.

GraphCMS

GraphCMS is a fully-hosted SaaS platform that's used by over 30,000 teams of all sizes across the world. Their queries are cached across 190 edge CDN nodes globally, meaning that wherever you're located, pulling data from GraphCMS into your Gatsby projects should be blazingly fast. Let's introduce ourselves to using GraphCMS by creating a list of hobbies within the tool that we can then ingest within our application:

1. Navigate to the GraphCMS website (`graphcms.com`) and log in.
2. Create a new blank project and pick the region you want to host your data in.
3. Navigate to your project's **schema** and create a hobby **model**. The schema is the blueprint for your data graph. Your schema is built from the models you create, the fields they contain, and their relationships. Clicking the **add** button next to `Models` will open the following dialog:

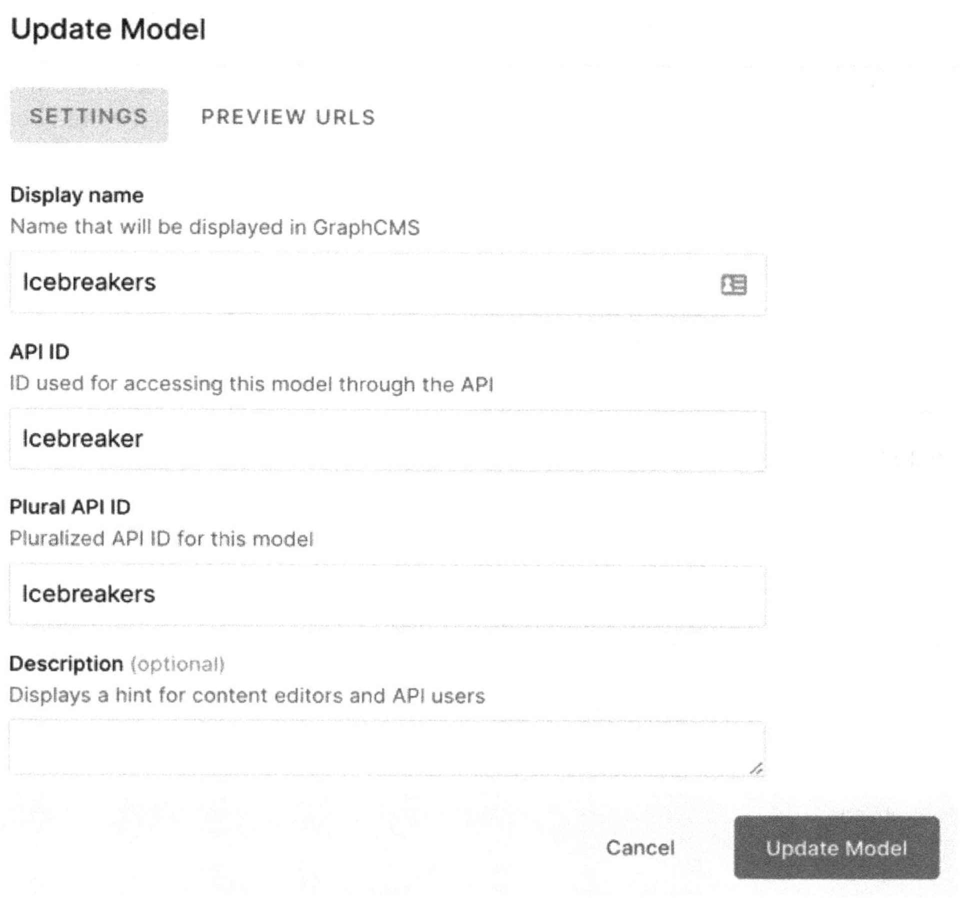

Update Model

SETTINGS   PREVIEW URLS

Display name
Name that will be displayed in GraphCMS

> Icebreakers

API ID
ID used for accessing this model through the API

> Icebreaker

Plural API ID
Pluralized API ID for this model

> Icebreakers

Description (optional)
Displays a hint for content editors and API users

Cancel   Update Model

Figure 3.2 – Model creation in GraphCMS

Here, you can see I am creating a model called **Icebreakers**. You'll notice that you need to provide an **API ID** and its plural form to make it easier to distinguish between when you are querying a single item versus the whole collection. Upon hitting **Update Model**, you should see that **Icebreakers** has been added to the model on the left sidebar.

4. We can now start to define what type of data is in our Icebreakers model by adding fields. Upon clicking on the Icebreakers model, you will see many field options on the right-hand side. We can use these to explain to GraphCMS what format our data will take. In our case, a hobby consists of one to three words each, so it would be appropriate to use the **Single Line Text** field option. Selecting this option will open the following dialog:

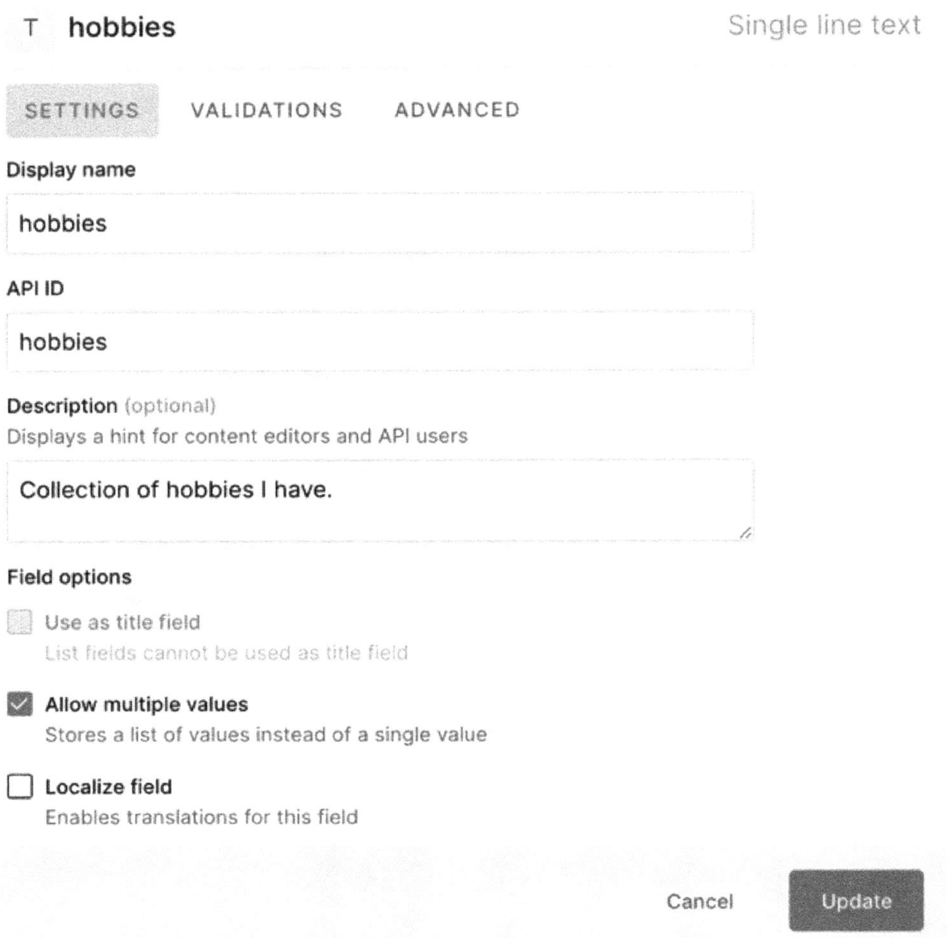

T   **hobbies**   Single line text

SETTINGS   VALIDATIONS   ADVANCED

Display name

hobbies

API ID

hobbies

Description (optional)
Displays a hint for content editors and API users

Collection of hobbies I have.

Field options

☐ Use as title field
   List fields cannot be used as title field

☑ **Allow multiple values**
   Stores a list of values instead of a single value

☐ **Localize field**
   Enables translations for this field

Cancel   Update

Figure 3.3 – Field creation in GraphCMS

Enter an appropriate display name and API ID, such as **hobbies**. Write **Collection of hobbies I have** as the description. I have also checked **Allow multiple values** so that we can store a list of hobbies instead of one. Click **Update** to save this configuration.

5. Navigate to the content section of the site. Click **Create item** at the top right of the page. This will open the following window:

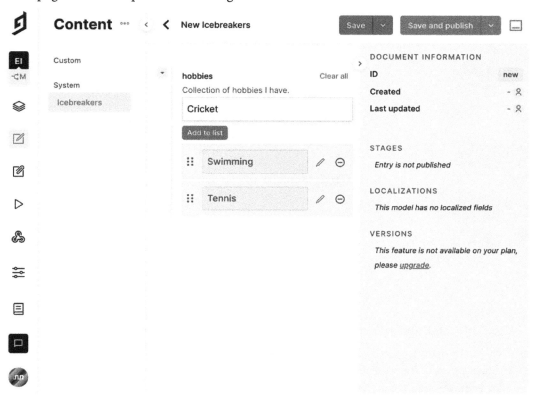

Figure 3.4 – Populating content in GraphCMS

We can now start to fill in our hobbies, adding them to the list as we go. Once you've done this, hit **Save** at the top right.

6. Returning to the content window, you will see that your created icebreaker is in **Draft** mode. This means that we are not happy with the content yet and that we will not be able to retrieve it from the API yet:

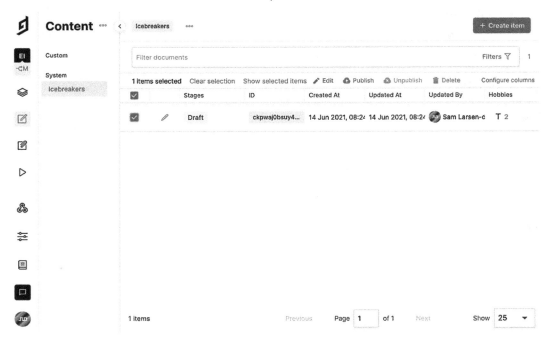

Figure 3.5 – GraphCMS content and its draft status

7. To make the content live, we need to publish it by selecting the item and then clicking the **Publish** button.

8. Next, we need to modify the endpoint settings to allow for public API access. By default, your GraphCMS API is not accessible from outside of their platform. You can change the settings for your public API access or create permanent authentication tokens with access permissions. Often, I lean toward keeping my data public as it is still only retrievable if you know the API's URL. Since it can't be edited by default, all of it will be displayed publicly on my site anyway.

Navigate to **Settings**, then **API Access**, and modify your public API permissions to the following:

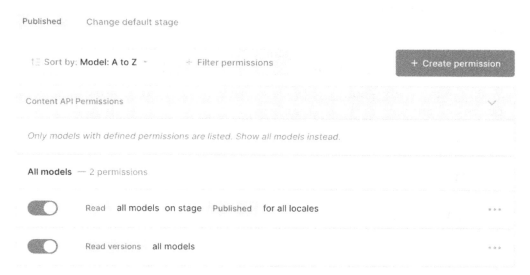

Figure 3.6 – GraphCMS public API settings

You will see that I have checked **Content from stage Published**. By doing so, we can now retrieve data that has been published via the URL endpoint, located at the top of the API's **Access** page.

9. Scroll to the top of this page and take note of your master URL endpoint. We will now move over to our Gatsby project and start ingesting data using this URL.

10. Open a terminal at the root of your project and install the necessary dependencies, the official GraphCMS source plugin, and dot-env:

```
npm install gatsby-source-graphcms gatsby-plugin-image
dotenv
```

gatsby-source-graphcms will allow us to source data from GraphCMS within our application, while dotenv is a zero-dependency module that loads environment variables from a .env file. We will be storing our API endpoint in the .env format. This plugin also requires gatsby-plugin-image under the hood, so make sure to install it. We will talk more about gatsby-plugin-image in *Chapter 5, Working with Images*.

11. Create a .env file at the root of your project and add your master URL endpoint for GraphCMS as a variable:

```
GRAPHCMS_ENDPOINT=Your-master-endpoint-url-here
```

This .env file is used to house environment variables. Be sure to replace the highlight with your master URL endpoint from *Step 6*. This file should not be committed to source control and, as such, should be added to your .gitignore.

12. Modify your gatsby-config.js file so that it includes gatsby-plugin-image and gatsby-source-graphcms:

```
require("dotenv").config()

module.exports = {
  ...
  plugins: [
    ...
    'gatsby-plugin-image',
    {
      resolve: 'gatsby-source-graphcms',
      options: {
        endpoint: process.env.GRAPHCMS_ENDPOINT,
      },
    },
    ...
  ],
};
```

Firstly, we use dotenv to load in our create .env file, and then we use that variable within the plugin configuration of gatsby-source-graphcms.

13. We can now start our development server. You will notice that when the development server starts, a new folder is created called `graphcms-fragments`. This folder is maintained by the plugin and contains fragments that explain the structure of our data to the GraphQL data layer.

14. At this point, we can query our data as if it were any other source. First, we must construct a query:

```
query Hobbies {
  graphCmsIcebreaker {
    hobbies
  }
}
```

Here, I have created a query that extracts our hobbies array from the auto-generated `graphCmsIcebreaker` source.

15. We can now embed this query in our `about` page:

```
import React from "react";
import { graphql } from "gatsby";
import Layout from "../components/layout/Layout";
import { MDXRenderer } from "gatsby-plugin-mdx";

export default function About({ data }) {
  const {
    mdx: { body },
    graphCmsIcebreaker: { hobbies },
  } = data;
  return (
    <Layout>
      <div className="max-w-5xl mx-auto py--16 lg:py-24
        text-center">
        <MDXRenderer>{body}</MDXRenderer>
        <div>
          <h2>Hobbies</h2>
          {hobbies.join(", ")}
        </div>
      </div>
    </Layout>
```

```
    );
  }

export const query = graphql'
  {
    mdx(frontmatter: { type: { eq: "bio" } }) {
      body
    }
    graphCmsIcebreaker {
      hobbies
    }
  }
';
```

You'll notice that I have just appended the new query to the existing page query, bundled into the same GraphQL string. Gatsby expects to only find one query per page. I then deconstructed the data prop to retrieve the hobbies array.

Now that we understand how GraphCMS works, let's turn our attention to how you would implement one of GraphCMS's competitors, Prismic.

Prismic

Prismic is smaller than GraphCMS, with around 5,000 paying customers. One feature that makes it stand out is they offer **dynamic multi-session previews**, allowing you to share multiple simultaneous dynamic previews (with shareable links) in Gatsby. This can improve your workflow when you're working with clients and you need to send the client's site content back and forth. Let's learn how to integrate Prismic by adding a list of hobbies within the UI so that we can then ingest them within our Gatsby site:

1. Create a folder in /src called schemas. Unlike GraphCMS, Prismic does not automatically create the schemas for us; instead, we will retrieve them using the Prismic UI as we create them.

2. Navigate to Prismic's website (prismic.io) and log in. Create a new repository with the free plan (you can always scale up later if you need to).

3. Click the **Create your first custom type** button and select the **single** type. Name your type **Icebreaker** and submit.

4. Scroll to the bottom of the build-mode sidebar on the right and drag a group into the central page:

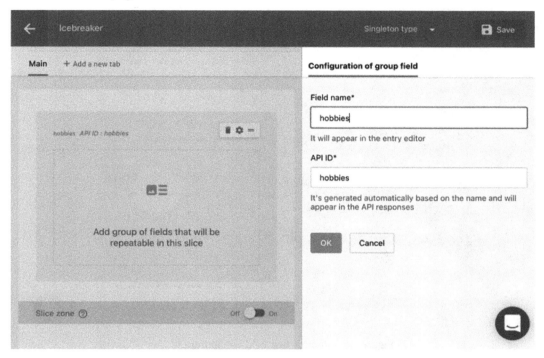

Figure 3.7 – Prismic group field options

5. Name your field **hobbies**; the corresponding API ID should populate on its own. Click **OK** to confirm this.

6. Drag a rich text field into this group:

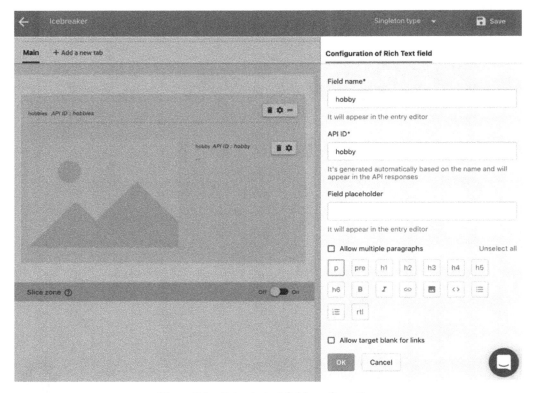

Figure 3.8 – Prismic text field configuration

This will open the side panel shown to the left of the preceding screenshot. We will use the rich text field as the type for a single hobby. First, let's give it a name – **hobby** seems appropriate. Ensure that **API ID** matches the assigned name. Uncheck the **Allow multiple paragraphs** box and then ensure that only the paragraph object is highlighted. By doing so, we can ensure that our hobbies are always single lines that only consist of paragraphs. Submit this using the **OK** button.

7. Save the document.

8. Now that we have defined our type, navigate to the JSON editor and copy its contents.

9. Create a new file inside your `schemas` folder called `icebreaker.json` and paste the JSON you have copied.

10. Navigate back home and click on **Documents.** Then click the *pencil icon* button to create a new instance of your Icebreaker type:

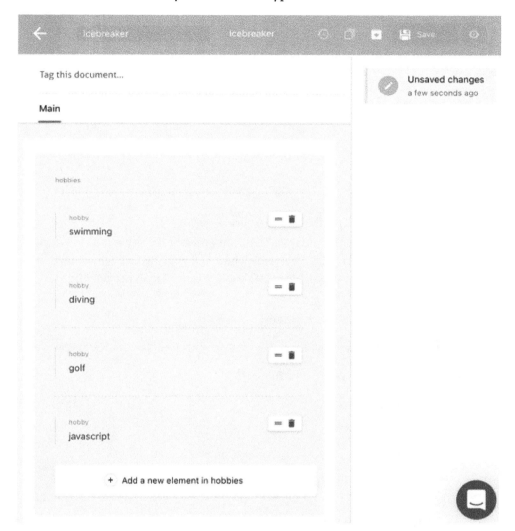

Figure 3.9 – Prismic collection interface

You can now use your hobbies type to create your data. Once you are happy with your list of hobbies, you can hit **Save**, followed by **Publish**.

11. Return home, navigate to **Settings**, and click on **API and security**. Ensure that your repository security is set to **Public API for Master only**:

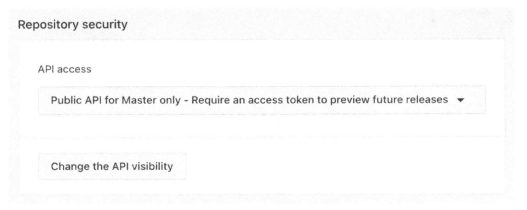

Figure 3.10 – Repository security

This means that anyone with your API URL can access what is currently live but not preview future releases. Make a note of your API entry point, which should be located at the top of this page. Now, let's look at our Gatsby project and start ingesting data using that URL.

12. Install the Gatsby Prismic source plugin:

```
npm install gatsby-source-prismic gatsby-plugin-image
```

13. Modify your `gatsby-config.js` file:

```
module.exports = {
  ...
  plugins: [
    ...
    'gatsby-plugin-image',
    {
      resolve: 'gatsby-source-prismic',
      options: {
        repositoryName: 'elevating-gatsby',
        schemas: {
          icebreaker:
            require('./src/schemas/icebreaker.json'),
        },
      }
```

```
    },
    ...
  ],
};
```

Here, you are adding the source plugin for Prismic to your Gatsby configuration. Be sure to change the repository name to that of your site. If you're unsure what your repository's name is, you can find it in your API URL. We are also directing the plugin to use the schema we have created for our icebreaker. The plugin is also dependent on `gatsby-plugin-image`, so make sure it has been added to your configuration.

14. We can now start our development server and query our data as normal. Upon opening GraphiQL, you should see `prismicIcebreaker` as a new source that we can use to query for our hobbies:

```
query Hobbies {
  prismicIcebreaker {
    data {
      hobbies {
        hobby {
          text
        }
      }
    }
  }
}
```

Here, we are retrieving the text value of every hobby from within the `hobbies` object.

15. We can now embed this query in our `about` page:

```
import React from "react";
import { graphql } from "gatsby";
import Layout from "../components/layout/Layout";
import { MDXRenderer } from "gatsby-plugin-mdx";

export default function About({ data }) {
  const {
    mdx: { body },
```

```
        prismicIcebreaker: {
          data: { hobbies },
        },
      } = data;
      return (
        <Layout>
          <div className="max-w-5xl mx-auto py-16 lg:py-24
            text-center">
            <MDXRenderer>{body}</MDXRenderer>
            <div>
              <h2>Hobbies</h2>
              <ul>
                {hobbies.map(({ hobby }) => (
                  <li>{hobby.text}</li>
                ))}
              </ul>
            </div>
          </div>
        </Layout>
      );
    }


    export const query = graphql'
      {
        mdx(frontmatter: { type: { eq: "bio" } }) {
          body
        }
        prismicIcebreaker {
          data {
            hobbies {
              hobby {
                text
              }
            }
          }
        }
```

```
    }
';
```

As we did when we looked at GraphCMS, I have just appended the new query to the existing page query. Our data is then passed in as the `data` prop and is available for us to use in whatever way we wish.

You should be starting to see the power of using GraphQL in Gatsby. As soon as we have ingested data, we can use the same format to query it every time. Using these two as examples, you should feel comfortable sourcing data from another CMS using a source plugin.

Summary

In this chapter, you learned how to use Gatsby's data layer. You learned about the basics of how to explore your GraphQL data layer via GraphiQL and should now feel comfortable sourcing and ingesting data into your Gatsby project from a multitude of different sources – `siteMetadata`, Markdown, MDX, and CMSes using their plugins. If you are interested in how source plugins are created and how to make your own, check out *Chapter 10*, *Creating Gatsby Plugins*.

In the next chapter, we will create and use reusable templates for pages that appear more than once, such as blog pages. This is great for when you have multiple pieces of data that you want to make use of while using the same layout.

4
Creating Reusable Templates

This chapter is where you will really begin to see the power that Gatsby brings to larger sites. You will learn about how we can programmatically create pages using reusable templates and data sourced via GraphQL. By the end of this chapter, you will have created lists of blog posts, blog pages, and tag pages. You'll also understand how to introduce pagination and search functionality to your site.

All the pages we have created up until now have been single instances, meaning there is only one copy of that page on the site (for example, our index page, of which there will be only one copy ever). But what happens when we consider pages such as blog pages? It would be a very laborious process to create a single instance page for each post. So, instead, we can use **templates**. A template is a multi-instance of a page component that is mapped to data. For every node in a GraphQL query, we can create a page using this template and populate it with the data of that node.

Now that we understand what we mean by templates in Gatsby, let's create our first few templates, and then **programmatically** create pages with them.

In this chapter, we will cover the following topics:

- Defining templates
- Creating templates and programmatic page generation
- Search functionality

Technical requirements

To complete this chapter, you will need to have completed *Chapter 3*, *Sourcing and Querying Data (from Anywhere!)*. You'll get the most out of this chapter if you have a collection of blog posts that we can use to build our pages, ingested into Gatsby. The source doesn't matter – you'll be ready to start this chapter if you can see them in your GraphQL data layer. If you don't have any posts to hand, you can find some placeholder Markdown files that you can ingest in Gatsby here: `https://github.com/PacktPublishing/Elevating-React-Web-Development-with-Gatsby-4/tree/main/Chapter04/placeholder-markdown`.

The code for this chapter can be found at `https://github.com/PacktPublishing/Elevating-React-Web-Development-with-Gatsby-4/tree/main/Chapter04`.

> **Important Note**
>
> To keep the code snippets at a manageable size, many of the examples in this chapter have styling omitted and comments pointing to code we've already written. To see a fully styled version of these components, please navigate to this book's code repository.

Creating templates and programmatic page generation

In this section, we will programmatically generate pages using templates. We will be creating blog pages, blog list preview pages, and tag pages. To make all of this work correctly, it is important to ensure that each node of data that you are ingesting to populate blog pages contains the following:

- **Title**: The title of the blog post.
- **Description**: A one-line description of what the blog post contains.

- **Date**: The date that the post should be published.
- **Tags**: A list of tags that the blog post is associated with.
- **Body**: The main content of the post.

If you are sourcing more than one type of content from the same source, it would be a good idea to also include a **type** field. This will allow you to filter out nodes that don't belong to this type.

The method for adding these to your nodes will change, depending on the source. However, in the case of Markdown, you could create your posts in the following format:

```
---
type: Blog
title: My First Hackathon Experience
desc: This post is all about my learnings from my first
  hackathon experience in London.
date: 2020-06-20
tags: [hackathon, webdev, ux]
---
# Body Content
```

Here, we added `title`, `desc`, `date`, and `tags` to `frontmatter`. Our body content would then be everything following `frontmatter`.

> **Important Note**
>
> I will be querying data from local Markdown files within this chapter. If you are sourcing content from another type of local or a remote source, you can still use all the code except the queries and node field manipulation, which you will have to modify to work with your source. If you are struggling to construct your queries, refer back to *Chapter 3, Sourcing and Querying Data (from Anywhere!)*.

Regardless of your source, you should ensure that your content is populated with the same fields to ensure GraphQL queries for blog-related data are always consistent.

Now that we have established the necessary blog node data fields, let's create blog post pages using our data.

Blog post template

In this section, we will create pages for each blog post we have. We will be creating and using our first template to do this with the help of the following steps:

1. Modify your `gatsby-node.js` file so that it includes the following code:

```
const { createFilePath } = require('gatsby-source-
  filesystem');
exports.onCreateNode = ({ node, getNode, actions }) => {
  const { createNodeField } = actions;
  if (node.internal.type === 'MarkdownRemark') {
    const slug = createFilePath({ node, getNode,
    basePath: 'pages' });
    createNodeField({
      node,
      name: 'slug',
      value: slug,
    });
  }
};
```

The `onCreateNode` function is called whenever a new node is created. Using this function, we can transform nodes by adding, removing, or manipulating their fields. In this specific case, we are adding a `slug` field if the node is of the `MarkdownRemark` type. A `slug` is an address for a specific page on our site, so in the case of our blog page, we want every blog post to have a unique `slug` where it will render on the site. Creating slugs from filenames can be complicated as you need to handle characters that would break URL formatting. Luckily, the `gatsby-source-filesystem` plugin ships with a function called `createFilePath` for creating them.

2. Verify that each blog page has a `slug` by running your development server and using GraphiQL to explore your nodes. If you are using Markdown, you should find it within the `fields` object on `MarkdownRemark` nodes.

3. Create a new folder inside `src` called `templates` to house our page templates.

4. Create a new file inside `templates` called `blog-page.js`. This is the file where we will create our blog page template.

5. Add the following code to the `blog-page.js` file:

```
import React from "react";
import Layout from "../components/layout/Layout";
import TagList from "../components/blog-posts/TagList"

export default function BlogPage() {
  return (
    <Layout>
      <div className="max-w-5xl space-y-4 mx-auto
        py-6 md:py-12 overflow-x-hidden lg:overflow-x-
        visible">
        <h1 className="text-4xl font-bold">Blog
          Title</h1>
        <div className="flex items-center space-x-2">
          <p className="text-lg opacity-50">Date</p>
          <TagList tags={["ux"]} />
        </div>
        <div>
          Article Body
        </div>
      </div>
    </Layout>
  );
}
```

Here, we are creating the blog post template with a static set of data that we will switch out for real content shortly. You can see that we have a heading containing our blog post title. We then follow it with the blog's `Date` and a `TagList` component, which we will make shortly. Finally, we have the main `Article Body`.

6. Create a folder inside `src/components` called `blog-posts`, in which we will store any component related to the blog.

7. Create a `TagList` component in the `src/components/blog-posts` file. We will use this component whenever we want to render a list of `tag` badges on the screen:

```
import React, { Fragment } from "react";

const TagList = ({ tags }) => {
  return (
    <Fragment>
      {tags.map((tag) => (
        <div
          key={tag}
          className="rounded-full px-2 py-1 uppercase
            text-xs bg-blue-600 text-white"
        >
          <p>{tag}</p>
        </div>
      ))}
    </Fragment>
  );
};

export default TagList
```

This component takes in an array of `tags` as a prop, maps over them, and returns a styled `div` containing that `tag`. This is all wrapped up in a `Fragment` component. By using a `Fragment`, we can avoid enforcing the ordering and positioning of our `tags`, and can instead allow the parent element to decide.

Now that we have created a template file and its components, we can use it within our `gatsby-node.js` file.

8. Add the following code to the top of your `gatsby-node.js` file:

```
const path = require('path');
const { createFilePath } = require('gatsby-source-
  filesystem');
exports.createPages = async ({ actions, graphql,
  reporter }) => {
  const { createPage } = actions;
```

```js
const BlogPostTemplate =
  path.resolve('./src/templates/blog-page.js');
const BlogPostQuery = await graphql('
  {
    allMarkdownRemark(filter: { frontmatter: { type:
      { eq: "Blog" } } }) {
      nodes {
        fields {
          slug
        }
      }
    }
  }
');
if (BlogPostQuery.errors) {
  reporter.panicOnBuild('Error while running GraphQL
    query.');
  return;
}
BlogPostQuery.data.allMarkdownRemark.nodes.forEach(({
fields: { slug } }) => {
  createPage({
    path: 'blog${slug}',
    component: BlogPostTemplate,
    context: {
      slug: slug,
    },
  });
});
};
```

9. Here, we are utilizing the `createPages` function, which allows us to create pages dynamically. To ensure you can query all your data, this function is only run once all your data has been sourced. Inside this function, we first destructure the `actions` object to retrieve the `createPage` function. Then, we tell Gatsby where to find our blog post template. With these two pieces in place, we are now ready to query our data. You should see a familiar GraphQL query for selecting the `slug` from all the Markdown where the type is *Blog*. We then have a small `if` statement to catch errors but, assuming it's successful, we then have all the data we need to create pages. We can loop through the result of our data and loop through every data node, creating a page for each one by specifying a path (using `slug`) and our template. You'll also notice that we are defining some `context` here. Data that's defined in `context` is available in page queries as GraphQL variables, which will make it easy to map the correct Markdown content to the correct pages in the following steps. Restart your development server and open the development 404 page by navigating to any non-existent route on the port. This will display a list of pages on your site, including the pages we have just created. Clicking on one should render the static content we defined when creating the template. Now that these pages have been created successfully, let's navigate back to the template and modify it to retrieve the correct content instead of the static content.

10. Modify the `src/templates/blog-post.js` file with the following code:

```
import React from "react";
import { graphql } from "gatsby";
import Layout from "../components/layout/Layout";
import TagList from "../components/blog-posts/TagList"
export default function BlogPage({data}) {
  const {blogpost: {frontmatter: {date, tags, title},
    html}} = data
  const shortDate = date.split("T")[0]
  return (
    <Layout>
      <div className="max-w-5xl space-y-4 mx-auto
        py-6 md:py-12 overflow-x-hidden lg:overflow-x-
        visible">
        <h1 className="text-4xl font-
          bold">{title}</h1>
        <div className="flex items-center space-x-2">
          <p className="text-lg opacity-
            50">{shortDate}</p>
```

```
            <TagList tags={tags} />
          </div>
          <div className="prose max-w-5xl"
            dangerouslySetInnerHTML={{__html:html}}/>
        </div>
      </Layout>
    );
  }
export const pageQuery = graphql'
query($slug: String!) {
    blogpost: markdownRemark(fields: {slug: {eq:
      $slug}}) {
        frontmatter {
          date
          title
          tags
        }
        html
      }
    }'
```

You will see that our template has a page query appended to it; this is where the slug property we defined in the gatsby-node.js file comes in handy. We can use that slug to find the blog post where slug matches in the node's fields. We query for all the data that we need to populate this page with and retrieve date, title, tags, and the Markdown HTML. This is then passed into the template via the data prop, exactly like in our single instance pages. We can then use this content to swap out the static placeholder content we had previously.

11. By restarting your development server and navigating to one of your blog pages again, you should now see it populated with its node data. You've successfully made your first programmatic pages!

As we only have a few blog posts, creating all of these pages won't take very long. However, what happens if you have thousands of pages to create? Instead of waiting for all your site pages to build, you can instruct Gatsby to defer the generation of some of these pages. You can do this by passing `defer:true` to the `createPage` function in `gatsby-node.js`, like so:

```
createPage({
    path: 'blog${slug}',
    component: BlogPostTemplate,
    defer: true,
    context: {
      slug: slug,
    },
});
```

With this change, any page that's created in this way will be built the first time that that page is requested instead of at build time. This feature changes the kind of build from a static build to a hybrid build. For more information on this difference, please read *Chapter 9, Deployment and Hosting*.

Now that we have created blog post pages, we must have some way of linking to them from our other pages. Let's create a blog preview template page, where we can have a list of our blog posts with previews and a link to the pages we have just created.

Blog preview template

While we could create a single list of blog posts and render it, it is a standard pattern for websites to divide lists of blog posts, articles, and products using **pagination**. Using pagination within your site has three main benefits:

- **Better page performance**: If every article includes an image in the preview, then with every added item, we are increasing the amount of data we need to transfer to the client significantly. By introducing pagination, the client will only download small segments of data as they browse a group of items. This leads to faster page load times, which is particularly important in areas with low bandwidth.

- **Improved user experience**: Displaying all the content on a single page could overwhelm the user, so instead, we must break down our content into small and manageable chunks.

- **Easier navigation**: If we render hundreds of products in one continuous list, the user will have no idea how many products are there while scrolling through. By breaking content down into multiple pages with a set quantity of products on each, the user can understand the scale of your content better.

With all that in mind, let's create a paginated blog preview page using a template:

1. Create a `Pagination` component in the `src/components/blog-posts` file:

```
import React from "react";
import { Link } from "gatsby";
const Pagination = ({ numPages, currentPage }) => {
 var pageArray = [];
 for (var i = 1; i <= numPages; i++) pageArray[i] = i;
  return (
    <div>
      <ul>
        {currentPage !== 1 && (
          <li>
            <Link to={currentPage === 2 ? '/blog' :
              '/blog/${currentPage - 1}'}>
              Previous
            </Link>
          </li>
        )}
        {pageArray.map((pageNum) => (
          <li key={'pageNum_${pageNum}'} >
            <Link to={pageNum === 1 ? '/blog' :
              '/blog/${pageNum}'}>
              {pageNum}
            </Link>
          </li>
        ))}
        {currentPage !== numPages && (
          <li>
            <Link to={'/blog/${currentPage +
              1}'}>Next</Link>
          </li>
```

```
        ) }
      </ul>
    </div>
  );
};
export default Pagination;
```

Here, we have created the component that will allow us to access paginated blog preview pages. The component contains the number of pages and the current page as props. Using these two pieces of information, we can determine whether a user can navigate forward or back from their current page. How this component works is best explained by seeing how it renders:

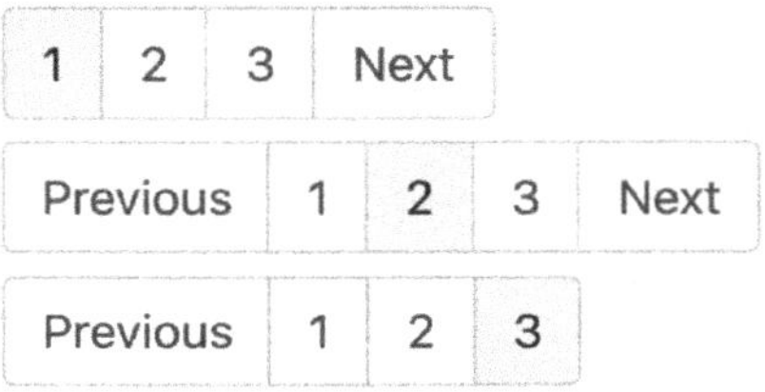

| 1 | 2 | 3 | Next |

| Previous | 1 | 2 | 3 | Next |

| Previous | 1 | 2 | 3 |

Figure 4.1 – Pagination component states

In the first case, the current page is **1**, so there is no need to render a **Previous** button. Instead, we only show the preceding pages and the **Next** button. In the second case, we are on page **2**, where the user can navigate both forward and back, and as such, we can render the **Previous** and **Next** buttons. In the last case, we are on the last page, so we don't need to render the **Next** button.

2. Create a new template in `src/templates/` called `blog-preview.js` and add the following page query:

```
/*
      Space for page component
*/
export const pageQuery = graphql'
  query($skip: Int!, $limit: Int!) {
    blogposts: allMarkdownRemark(
      limit: $limit
      skip: $skip
      filter: { frontmatter: { type: { eq: "Blog" } } }
      sort: { fields: frontmatter___date, order: DESC }
```

```
    ) {
        nodes {
            frontmatter {
                date
                title
                tags
                desc
            }
            fields {
                slug
            }
        }
    }
  ';
```

The query within this file sources data from `allMarkdownRemark` (which I have named `blogposts` in this query). The `blogposts` query retrieves all the Markdown where the `frontmatter` type is equal to `Blog`. It sorts the collection of posts by descending `date`. Here's where things get interesting – we also provide a `skip` and `limit` to the query. `skip` tells the query how many documents from the collection to skip over, while `limit` tells the query to limit the number of results to that quantity. We will be providing `skip` and `limit` to the page context, as well as `numPages` and `currentPage`, within our `gatsby-config.js` file.

3. Create the page component before the query in the `blog-preview.js` file:

```
import React from "react";
import { graphql, Link } from "gatsby";
import Layout from "../components/layout/Layout";
import Pagination from "../components/blog-
  posts/Pagination";
import TagList from "../components/blog-posts/TagList"

export default function BlogPreview({ pageContext,
  data }) {
  const {
    numPages,
    currentPage
```

```
  } = pageContext
  const {
    blogposts: { nodes },
  } = data;
  // return statement
}
```

As with our other queries, when the query at the end of this file runs, it will provide data to our page via the data prop. Here, we are destructuring pageContext to access numPages and currentPage. We are also using destructuring data to get nodes from the blogposts query. We will add our render via the return statement in the following step.

4. Create the return statement in this same file:

```
return (
  <Layout>
    <div className="max-w-5xl mx-auto space-y-8 py-6
    md:py-12">
      {nodes.map(
        ({ frontmatter: { date, tags, title, desc },
          fields: { slug } }) => (
          <div>
            <Link to={'/blog${slug}'}>
              <h2 className="text-2xl font-
                medium">{title}</h2>
              <div className="flex items-center
                space-x-2">
                <p className="text-lg opacity-
                  50">{date.split("T")[0]}</p>
                <TagList tags={tags}/>
              </div>
              <p>{desc}</p>
            </Link>
          </div>
        )
      )}
      <Pagination numPages={numPages}
```

```
                   currentPage={currentPage} />
       </div>
     </Layout>
  );
```

We use `nodes` from the two sources to map through posts, render a preview of each (making use of the `TagList` component), as well as render our `Pagination` component. Now that we have created our template, we can ingest it into our `gatsby-config.js` file.

5. Modify your `gatsby-config.js` file's `createPages` function with the following code:

```
exports.createPages = async ({ actions, graphql,
  reporter }) => {
  const { createPage } = actions;
  const BlogPostTemplate =
   path.resolve('./src/templates/blog-page.js');
  const BlogPreviewTemplate =
   path.resolve('./src/templates/blog-preview.js');
  // BlogPostQuery
  const BlogPosts =
    BlogPostQuery.data.allMarkdownRemark.nodes;
  const postsPerPage = 6;
  const numPages = Math.ceil(BlogPosts.length /
    postsPerPage);
  Array.from({ length: numPages }).forEach((_, i) => {
    createPage({
      path: i === 0 ? '/blog' : '/blog/${i + 1}',
      component: BlogPreviewTemplate,
      context: {
        limit: postsPerPage,
        skip: i * postsPerPage,
        numPages,
        currentPage: i + 1,
        slug: i === 0 ? '/blog' : '/blog/${i + 1}',
      },
    });
```

```
// Blog Post Page Creation
});
//
};
```

First, we require our new `BlogPreviewTemplate`, then we run our Markdown query as normal. As we will be now using `BlogPostQuery.data.allMarkdownRemark.nodes` in two places (blog previews and blog post page creation), we can assign it to a constant. We will also assign two more constants – the number of posts per page (`postsPerPage`) and the number of pages (`numPages`) that we will need for pagination. `postsPerPage` specifies how many posts we want on each of our paginated blog post previews. `numPages` calculates how many preview pages are needed by dividing the total number of posts by `postsPerPage` and then rounding up to the nearest whole integer using the `Math.ceil` function. We then create an `Array` with a `length` equal to the number of pages and loop through it using the `forEach` function. For each index (`i`), we use the `createPage` action. We provide this action with the path to where the page should be located, which is `/blog` if `i` is 0 and `/blog/i+1` for anything higher. We also provide `BlogPreviewTemplate` and `context`, which contain `limit` and `skip`, which we utilize on the page.

6. You are now ready to start your development server to verify that pagination is working. You should see your posts in descending date order located at `/blog`. If you have more posts than your `postsPerPage` value, you should also see your `Pagination` component, showing you that there are additional pages and allowing you to navigate there.

Now that we have implemented a blog preview page, let's use what we have learned to create one more collection of pages – tag pages.

Tag page template

As a user, seeing my posts in date order is not always enough – I may want to be able to find groups of posts associated with a single topic. Tag pages are pages you navigate to whenever you click on one of a blog post's tags. Navigating to one of these pages, you are presented with a list of posts that are associated with that tag.

Let's programmatically create tag pages for each tag that's present in our articles:

1. Install `lodash`:

```
npm i lodash
```

`lodash` is a JavaScript utility library that we will be using to make tags URL-friendly. Because a single tag might consist of multiple words, we need a way to remove the spaces. While you could create a function yourself to do this, `lodash` has a `.kebabCase()` function that works well for this use case.

2. Modify the `TagList` component to turn our `tag` badges into `Link` components:

```
import React, { Fragment } from "react";
import { Link } from "gatsby";
import { kebabCase } from "lodash"
const TagList = ({ tags }) => {
  return (
    <Fragment>
      {tags.map((tag) => (
        <Link key={tag}
            to={'/tags/${kebabCase(tag)}'}>
        <div
            key={tag}
            className="rounded-full px-2 py-1 uppercase
              text-xs bg-blue-600 text-white"
          >
          <p>{tag}</p>
        </div>
        </Link>
      ))}
    </Fragment>
  );
};
export default TagList
```

As `Link` components, they need a `to` prop. This prop should point to where your `tag` pages will be created – in our case, `/tags/tag-name` is the location. We can use the `kebabCase` function from `lodash` to ensure that any spaces in tags are turned into hyphens.

3. Create a `tags.js` file in the `src/templates` folder:

```
/*
      Space for page component
*/
```

```
export const pageQuery = graphql'
  query($tag: String) {
    blogposts: allMarkdownRemark(
      sort: { fields: [frontmatter___date], order:
        DESC }
      filter: { frontmatter: { tags: { in: [$tag] },
        type: { eq: "Blog" } } }
    ) {
      totalCount
      nodes {
        frontmatter {
          date
          title
          tags
          desc
        }
        fields {
          slug
        }
      }
    }
  }
';
```

4. This component will appear very similar to our previously constructed `blog-preview.js` file in the *Blog preview template* section, except for a minor change to the query. In this query, we still source our Markdown content, but this time, we filter out the posts that do not contain the page's tag.

5. Create the page component before the query in the `tags.js` file:

```
import React from "react";
import { graphql, Link } from "gatsby";
import Layout from "../components/layout/Layout";
import TagList from "../components/blog-
  posts/TagList";

export default function Tags({ pageContext, data }) {
```

```
const { tag } = pageContext;
const {
  blogposts: { nodes },
} = data;
return (
  <Layout>
    <div>
      <p>Posts tagged with "{tag}"</p>
      {nodes.map(
        ({ frontmatter: { date, tags, title, desc },
          fields: { slug } }) => (
          <div>
            <Link to={'/blog${slug}'}>
              <h2>{title}</h2>
              <div>
                <p>{date.split("T") [0]}</p>
                <TagList tags={tags} />
              </div>
              <p>{desc}</p>
            </Link>
          </div>
        )
      )}
    </div>
  </Layout>
);
}
```

The page then renders a paragraph containing the tag you are currently filtering posts with, followed by the filtered list of posts. Each post preview is rendered with its `title`, `date`, description (`desc`), and `tags`, just like in the `blog-preview.js` file.

> **Important Note**
>
> If you intend to render the same items in the lists of both your `blog-preview.js` and `tags.js` files, then you should probably abstract the item preview component into a separate component. To keep these examples independent, I will not do this here.

6. Import `lodash` into the top of your `gatsby-config.js` file, next to the other imports:

```
const _ = require("lodash");
```

We will need to use lodash's `kebabCase` in this file as well.

7. Add your tag template and query to your `gatsby-config.js` files' `createPages` function:

```
exports.createPages = async ({ actions, graphql,
    reporter }) => {
// actions destructure & other templates
  const TagsTemplate =
    path.resolve('./src/templates/tags.js');
  const BlogPostQuery = await graphql('
  {
    allMarkdownRemark(filter: {frontmatter: {type:
    {eq: "Blog"}}}) {
      nodes {
        fields {
          slug
        }
      }
    }
    tagsGroup: allMarkdownRemark(filter: {frontmatter:
    {type: {eq: "Blog"}}}) {
      group(field: frontmatter___tags) {
        tag: fieldValue
      }
    }
  }');
// Error Handling, Blog Preview & Blog Post page
  creation
  BlogPostQuery.data.tagsGroup.group.forEach((group)
  => {
    createPage({
      path: 'tags/${_.kebabCase(group.tag)}/',
      component: TagsTemplate,
```

```
        context: {
            tag: group.tag,
        },
    });
  });
};
```

First, we acquire our new `TagsTemplate`. Then, we append our query with a new query to our Markdown source. This group (which we've named `tagsGroup`) retrieves an array containing every unique `tag` that is within `frontmatter` of our posts.

We can then use this new data to loop through every `tag` and create a `tag` page for each one. We pass a `path` to each `createPages` function, pointing to `tags/`, followed by the `tag` name that's parsed through the `kebabCase` function. We pass the `component` property we want it to build the page with, which in our case is `TagsTemplate`, at the beginning of this file. You will also notice that we are also passing `tag` to the page's `context` so that the page knows which `tag` it relates to.

8. You are now ready to start your development server to verify that the tag pages are working. Navigate to the development 404 page; you should see a page starting with `tags/` for each tag. Clicking on one of these, you should be presented with our tag page template and a list of blog posts associated with that tag.

> **Further Exercise**
> We've learned how to paginate blog lists, as well as create tag pages. Why not take this one step further and paginate your tag pages?

With that, we have learned how to programmatically create pages for blog posts, blog lists, and tags. Now, let's turn our attention to how we might create a site search so that as the site expands, finding our blog's content is easier.

Search functionality

There are many different ways of integrating a site search. Many options are both hosted and local. For small projects, such as the site we are creating, it's often better to opt for a local index solution as the number of pages you are searching through is never that large. This also means that your site search will work in offline scenarios, which can be a real plus.

Elasticlunr (`http://elasticlunr.com/`) is a lightweight full-text search engine in JavaScript for browser and offline search. Using the `elasticlunr` Gatsby plugin, content is indexed and then made available via GraphQL to **rehydrate** in an `elasticlunr` index. Search queries can then be made against this index to retrieve page information.

Let's integrate a site search using `elasticlunr`:

1. Install the `elasticlunr` Gatsby plugin:

    ```
    npm install @gatsby-contrib/gatsby-plugin-elasticlunr-
    search
    ```

2. Add the `elasticlunr` plugin to your `gatsby-config.js` plugins array:

    ```
    {
        resolve: '@gatsby-contrib/gatsby-plugin-
          elasticlunr-search',
        options: {
          fields: ['title', 'tags', 'desc'],
          resolvers: {
            MarkdownRemark: {
              title: node => node.frontmatter.title,
              tags: node => node.frontmatter.tags,
              desc: node => node.frontmatter.desc,
              path: node => '/blog'+node.fields.slug,
            },
          },
          filter: (node, getNode) =>
          node.frontmatter.type === "Blog",
        },
    },
    ```

As part of `options`, we provide the plugin with a list of `fields` that we would like to index. Then, we give it a `resolvers` object, which explains how to resolve `fields` for a source. Within our blog posts, we can retrieve `title`, `tags`, and `desc` from `frontmatter`. We can construct `path` with that specific content with a string and the data's `slug`. Finally, we also pass a `filter`. This `filter` tells the plugin to only use nodes where the `frontmatter` type is of the `Blog` type, as it is only our blog pages that we want to be searchable at this moment.

3. Create a `Search.js` component in the `src/layout` folder:

```
import React, { useState, useEffect } from "react";
import { Link } from "gatsby";
import { Index } from "elasticlunr";

const Search = ({ searchIndex }) => {
  const [query, setQuery] = useState("");
  let [index, setIndex] = useState();
  let [results, setResults] = useState([]);

  useEffect(() => {
    setIndex(Index.load(searchIndex));
  }, [searchIndex]);
};
export default Search
```

The `Search` component takes in `searchIndex` as a prop. One of the first things you will notice is a `useEffect` hook that loads the index into a state hook. Once we have loaded in the index, we can query it.

4. Create a `search` function below `useEffect`:

```
const search = (evt) => {
    const query = evt.target.value;
    setQuery(query);
    setResults(
      index
        .search(query, { expand: query.length > 2 })
        .map(({ ref }) =>
      index.documentStore.getDoc(ref))
    );
  };
```

You will see that whenever the `search` function is called, we `search` the index using our `query` string. You will notice that we are passing in `expand: query. length > 2` as an option to the `search` function. This tells `elasticlunr` to allow partial matches if more than two characters have been entered. If you allow partial matches for fewer characters, you will often find that you get an abundance of results that are not related to what the user is looking for. Once we have searched the index, we can `map` over `documentStore` within `index` and return the document results, which are then passed to state via the `useState` hook.

5.   Create the `search` result render function:

```
const searchResultSize = 3;
return (
    <div className="relative w--64 text-gray-600">
      <input
        type="search"
        name="search"
        placeholder="Search"
        autoComplete="off"
        aria-label="Search"
        onChange={search}
        value={query}
      />
      {results.length > 0 && (
        <div>
          {results
            .slice(0, searchResultSize)
            .map(({ title, description, path }) => (
              <Link key={path} to={path}>
                <p>{title}</p>
                <p className="text-
                xs">{description}</p>
              </Link>
            ))}
          {results.length > searchResultSize && (
            <Link to={`/search?q=${query}`}>
              <p>+ {results.length - searchResultSize}
                more</p>
```

```
            </Link>
        )}
      </div>
    )}
  </div>
);
```

We map over results from the state using the `results` value from the `useState` hook and render the results to the screen within our render function. For a slightly better user experience, it's often a good idea to include a `searchResultSize` constant. This value determines the maximum number of results to display. This stops cases where you have hundreds of results and see the overlay run off the page. Instead, if there are more results, we simply indicate to the user how many more results there are.

6. Modify your `Header.js` file to retrieve the site index and pass it to your `Search` component:

```
import React from "react";
import { Link, StaticQuery, graphql } from "gatsby";
import Search from "./Search";

const Header = () => (
  <header className="px-2 border-b w-full max-w-7xl
    mx-auto py-4 flex items-center justify-between">
    <Link to="/">
      <div className="flex items-center space-x-2
        hover:text-blue-600">
        <p className="font-bold text-2xl">Site
          Header</p>
      </div>
    </Link>
    <StaticQuery
    query={graphql'
      query SearchIndexQuery {
        siteSearchIndex {
          index
        }
      }
```

```
    '}
    render={data => (
      <Search
        searchIndex={data.siteSearchIndex.index}/>
    )} />
  </header>
);

export default Header;
```

Because `Header.js` is not a page component, we cannot append the `graphql` query to the end of the page as Gatsby is not looking for it. However, we can still locate data with the component by using `StaticQuery`. Static queries differ from page queries as they cannot accept variables like our pages can via page context. In this scenario, that's not a constraint as the search `index` is always static.

`StaticQuery` has two important props – `query` and `render`. `query` accepts a `graphql` query, while `render` tells the component what to render with the data from that query. In this particular instance, we are querying for the elasticlunr `index`, and then rendering our `Search` component using that `data`, passing `index` as a prop.

7. Now that we have completed our search functionality, restart your development server. You should see that the header of our site now contains our `Search` component. Try typing in a few characters and click on one of the results. You should be navigated to the corresponding page.

By adjusting the resolvers and using the same methodology and tools outlined here, we could add pages of different types to our results to create a true site-wide search.

Summary

In this chapter, you learned how to programmatically create pages using reusable templates. You should feel confident that you can now create pages using any GraphQL data source. We've implemented a list of blog posts with pagination, blog pages, tag pages, and created a site search for blog posts that even works offline.

In the next chapter, we will master the art of adding images to our Gatsby site. First, we will learn why importing images is not that simple, before creating images that progressively load in and are performant.

5
Working with Images

In this chapter, you will master the art of adding images to your Gatsby site. First, we will learn a little about the history of images on the web, before understanding why importing images is not as easy you might think. We will then move on to creating images that progressively load in and are performant.

In this chapter, we will cover the following topics:

- Images on the web
- The `StaticImage` component
- The `GatsbyImage` component
- Overriding the `gatsby-plugin-image` defaults
- Sourcing images from CMS

Technical requirements

To complete this chapter, you will need to have completed *Chapter 3, Sourcing and Querying Data (from Anywhere!).*

The code for this chapter can be found at `https://github.com/PacktPublishing/Elevating-React-Web-Development-with-Gatsby-4/tree/main/Chapter05`.

Images on the web

When was the last time you visited a website without any images? You might be thinking that this is a hard question to answer. Images are a critical part of websites and our browsing experience. We use images for logos, products, profiles, and marketing to convey information, entice, or excite through a visual medium. While images are great for these use cases (and many more!), they are the single largest contributor to page size. According to the HTTP Archive (`https://httparchive.org/`), the median page size on desktops is 2,124 KB. Images make up 973 KB of this size, which is roughly 45% of the total page size.

As images are so integral to our browsing experience, we cannot do away with them. But when they account for so much of the page size, we should be doing everything in our power to ensure that they are optimized, accessible, and as performant as possible. Newer versions of browsers (including Chrome) have **responsive image** capabilities built into them. Instead of providing a single source for the image, the browser can accept a source set. The browser uses this set to load an image at a different size, depending on the size of the device. This ensures the browser never loads images that are too big for the space available. Another way in which developers can optimize images, specifically in React, is with **lazy loading**. Lazy loading is the process of deferring the load of your images until a later point in time. This could be after the initial load, or when they specifically become visible on the screen. Using this technique, you can improve your site's speed, better utilize a device's resources, and reduce a user's data consumption.

Images in Gatsby

Manually creating high-performance sites that contain lazy-loaded images is a project in itself. Luckily, Gatsby has a plugin that takes the pain away when you're generating responsive, optimized images – `gatsby-plugin-image`.

This plugin contains two React image components with specific features aimed at creating a better user experience when using images, both for the developer and the site visitor:

- Loads the correct image for the device viewing the site.

- Reduces cumulative layout shift by holding the position of the image while it is loading.

- Uses a lazy loading strategy to speed up the initial load time of your site.

- Has multiple options for placeholder images that appear while the image is loading. You can make the images blur up or use a traced **Scalable Vector Graphics (SVG)** of the image in its place.

- Supports new image formats such as **WebP** if the browser can support it.

In the next section, we will begin looking at the first of the two image components in this chapter – the StaticImage component.

The StaticImage component

StaticImage is best used when an image will always remain the same. It could be your site logo, which is the same across all pages, or a profile photo that you use at the end of blog posts, or a home page hero section, or anywhere else where the image is not required to be dynamic.

Unlike most React components, the StaticImage component has some limitations on how you can pass the props to it. It will not accept and use any of its parents' props. If you are looking for this functionality, you will want to use the GatsbyImage component.

To get an understanding of how we utilize the StaticImage component, we will implement an image on the hero of our home page:

1. Create an assets folder next to your src folder. To keep things organized, it is good practice to keep images away from your source code. We will use the assets folder to house any visual assets.

2. Create a folder within assets called images. We will use this folder to store the visual assets of the image type.

3. Add an image to your images folder. The image file type must be in .png, .jpg, .jpeg, .webp, or .avif format.

4. Install gatsby-plugin-image and its dependencies:

    ```
    npm install gatsby-plugin-image gatsby-plugin-sharp
    gatsby-source-filesystem
    ```

    These dependencies spawn other node processes and may take a little longer to install compared to our previous installs.

5. Update your gatsby-config.js file so that it includes the three new dependencies:

    ```
    plugins: [
        'gatsby-plugin-image',
    ```

```
      'gatsby-plugin-sharp',
      'gatsby-transformer-sharp',
      // Other plugins
   ],
```

6. Import the StaticImage component into your index.js file:

```
import { StaticImage } from "gatsby-plugin-image";
```

7. Use the StaticImage component within the render on the page:

```
<StaticImage
    src="../../assets/images/sample-photo.jpeg"
/>
```

The src prop should be the relative path to the image from the current file, *not* the root directory.

8. Modify the image that's rendered via props:

```
<StaticImage
    src="../../assets/images/sample-photo.jpeg"
    alt="A man smiling"
    placeholder="tracedSVG"
    layout="fixed"
    width={400}
    height={600}
/>
```

Let's look at these props in detail:

- src: This prop is the relative path to the image from the current file.

- alt: As you would with a normal img tag, be sure to include an alt prop with text describing the image so that your image remains accessible.

- Placeholder: This prop tells the component what to display while the image is loading. Here, we have set it to tracedSVG, which uses a placeholder SVG (created by tracing the image) to fill the gap where the image will load in, but also gives the user a sense of the shape of whatever is in the photo. Other options include blurred, dominantColor, and none.

- `layout`: This prop determines the size of images that are produced as the output by the plugin and their resizing. Here, we have set it to `fixed` – this suggests that the image will be at a consistent size when it renders. Other layout options include `constrained`, which takes a maximum height or width and can scale down, and `fullWidth`, which will also resize to fit the container but is not restricted to the maximum height or width.

- `width` and `height`: We can use these props to specify the width and height of the image.

> **Tip**
>
> `StaticImage` can also take a remote source as its `src` prop. Any images that are specified as URLs will be downloaded at build time and then resized. Using remote images instead of local images is a great way to keep your repository small, but it should also be remembered that Gatsby does not know when that image changes if it is outside of your repository. If the image is changed on the remote server, it will only update when you rebuild your project.

Now that we understand how to utilize the `StaticImage` component, let's turn our attention to the other half of `gatsby-plugin-image` and learn about the `GatsbyImage` component.

The GatsbyImage component

If ever you need to use dynamic images, such as those embedded in your Markdown content, then you can use the `GatsbyImage` component.

Let's add hero images to our Markdown/MDX blog posts using the `GatsbyImage` component:

1. Install the `gatsby-transformer-sharp` npm package:

```
npm install gatsby-transformer-sharp
```

2. Add some images to `assets/images` that you would like to use as covers for your blog posts – one per blog post.

3. Update your `Gatsby-config.js` file so that it includes your `assets` source:

```
{
        resolve: 'gatsby-source-filesystem',
        options: {
```

```
            path: '${__dirname}/assets/images',
        },
    },
```

Unlike `StaticImage`, `GatsbyImage` requires that images are ingested into our data layer. We can use the `gatsby-source-filesystem` plugin to achieve this, but by giving it the path to our images.

4. For each blog post, modify the post file's `frontmatter` so that it includes a `hero` key that contains the relative path to the image:

```
---
type: Blog
title: My First Hackathon Experience
desc: This post is all about my learnings from my
  first hackathon experience in London.
date: 2020-06-20
hero: ../../assets/images/cover-1.jpeg
tags: [hackathon, webdev, ux]
---
```

Here, you can see an updated example of the `placeholder` Markdown with a `hero` key added. Be sure to replace the relative path in this example with the one to your image.

5. Add the `GatsbyImage` component and the `getImage` function as imports, at the top of the `src/templates/blog-page.js` file:

```
import { GatsbyImage, getImage } from "gatsby-plugin-image";
```

6. Modify the file's page query to reference the new images:

```
export const pageQuery = graphql'
  query($slug: String!) {
    blogpost: markdownRemark(fields: { slug: { eq:
      $slug } }) {
      frontmatter {
        date
        title
        tags
        hero {
```

```
            childImageSharp {
                gatsbyImageData(
                    width: 600
                    height: 400
                    placeholder: BLURRED
                )
            }
        }
      }
    html
    }
  }
';
```

You may notice that the data that's passed to the `gatsbyImageData` function looks very similar to the props of the `StaticImage` component that we saw in *Step 8* of the previous section. In this instance, we are using the `BLURRED` placeholder for the image, which uses a blurred, lower-resolution version of the image in place of the original image while it is loading. We are now able to retrieve the `hero` data from the component as it is included in the page query.

7. Use the new data within the `data` prop to get the image within the component:

```
const {
    blogpost: {
        frontmatter: { date, tags, title, hero },
        html,
    },
} = data;
const heroImage = getImage(hero)
```

First, retrieve `hero` from the `data` prop, and then use the `getImage` utility from `gatsby-plugin-image` to retrieve the image data that's required to render it and assign it to a `const`.

8. Render your image within your `return` statement:

```
<GatsbyImage image={heroImage} alt="Your alt text" />
```

Use the `const` defined in *Step 7* to render the image within a `GatsbyImage` component. Be sure to provide it with `alt` text to keep your image accessible – you could provide this via `frontmatter` as well if you wish.

9. Start or restart your development server and admire your hard work. Navigate to one of the blog posts and you should see your image blur in gracefully.

> **Further Exercise**
>
> We've learned how to add images to our blog pages, so why not use what you have learned to add a smaller version of the same image to the blog preview page? An implementation can be found in this book's GitHub repository (`https://github.com/PacktPublishing/Elevating-React-Web-Development-with-Gatsby-3/tree/main/Chapter05`) if you want to see how it can be achieved.

You might find yourself adding the same configuration to your images across the whole site. Let's find out how we can use the defaults to keep our code in **Don't Repeat Yourself** (**DRY**) form.

Overriding the gatsby-plugin-image defaults

To create a consistent look and feel, you may have included the same props with many instances of the two image components. Keeping these image components updated can be a monotonous task if your site is image-heavy. Instead, you could configure the defaults within the options of `gatsby-plugin-sharp`:

```
{
    resolve: 'gatsby-plugin-sharp',
    options: {
      defaults: {
        formats: ['auto', 'webp'],
        placeholder: 'blurred'
        quality: 70
        breakpoints: [300…]
        backgroundColor: 'transparent'
        tracedSVGOptions: {}
        blurredOptions: {}
        jpgOptions: {}
        pngOptions: {}
```

```
        webpOptions: {}
        avifOptions: {}
    }
  }
},
```

Let's look at each of these options in detail:

- `formats`: The file formats generated by the plugin.

- `placeholder`: Overrides the style of the temporary image.

- `quality`: The default image quality that's created.

- `breakpoints`: The widths to use for full-width images. It will never create an image with a width that's longer than the source.

- `backgroundColor`: The default background color of the images.

- `tracedSVGOptions` and `blurredOptions`: The default options to use for placeholder image types in the case that these are different from the global defaults.

- `jpgOptions`, `pngOptions`, `webpOptions`, and `avifOptions`: The default options to use for image types in the case that these are different from the global defaults.

We now have a good understanding of the `gatsby-plugin-image` package. There are, however, some important niches to using this package with other sources such as **Content Management Systems (CMSes)** – let's take a look.

Sourcing images from CMS

It is not always practical to store images within your repository. You may want someone other than yourself to be able to update or add images to your site without you needing to change the code. In these cases, serving images via CMS is preferable. It's still important that we use the Gatsby image plugin as we want our images to be performant, regardless of the source. To understand how we would integrate images via CMS, let's use an example. Let's add a profile image to our *about* page using `gatsby-plugin-image` and a CMS.

Important Note

Due to the vast number of headless CMSes in the market, we will continue to focus on the two mentioned in the *Sourcing data from a Headless CMS* section of *Chapter 3, Sourcing and Querying Data (from Anywhere!)*: GraphCMS and Prismic.

Both of the following sections will assume you have already installed the CMS's dependencies and integrated the CMS via your `gatsby-config.js` file. Please only implement one of the following.

Sourcing images from GraphCMS

By making some small modifications to our configuration and queries, we can ensure that we can source images from GraphCMS that utilize `gatsby-plugin-image` and load in on the site in the same way as those that are locally sourced:

1. Navigate to the GraphCMS website (`graphcms.com`) and log in.

2. Navigate to your project's assets and click the **upload** button.

3. Drag and drop a local image you wish to use onto the page. Be sure to take note of the file's name as we will need it later.

4. Publish the asset.

5. Modify your `gatsby-source-graphcms` plugin:

```
{
    resolve: 'gatsby-source-graphcms',
    options: {
      endpoint: process.env.GRAPHCMS_ENDPOINT,
      downloadLocalImages: true,
    },
  },
```

By modifying your `gatsby-source-graphcms` plugin's options to include the `downloadLocalImages` option, the plugin will download and cache the CMS's image assets within your Gatsby project.

6. Modify your about page's `query` so that it includes the `graphCmsAsset` source:

```
export const query = graphql`
  {
    markdownRemark(frontmatter: { type: { eq: "bio" } })
  {
      html
    }
    graphCmsAsset(fileName: { eq: "profile.jpeg" }) {
      localFile {
        childImageSharp {
```

```
        gatsbyImageData(layout: FULL_WIDTH)
      }
    }
  }
 }
';
```

As with the local queries we looked at in the *The GatsbyImage component* section, we are using `gatsbyImageData` within our `query` and can make use of any of the configuration options that it supports. Here, we are specifying that the image should be full width.

7. Add the `GatsbyImage` component and the `getImage` function as imports at the top of the `src/pages/about.js` file:

```
import { GatsbyImage, getImage } from "gatsby-plugin-
image";
```

8. Use the new data within the `data` prop to get the image within the component:

```
const {
    markdownRemark: { html },
    graphCmsAsset: { localFile },
  } = data;
  const profileImage = getImage(localFile);
```

First, retrieve `graphCmsAsset` from the `data` prop and then use the `getImage` utility from `gatsby-plugin-image` to retrieve the image data that's required to render it. Finally, assign it to a `const` called `profileImage`.

9. Render your image within your `return` statement:

```
return (
    <Layout>
      <div className="max-w-5xl mx-auto py-16 lg:py-24
        text-center">
        <GatsbyImage
          image={profileImage}
          alt="Your alt text"
          className="mx-auto max-w-sm"
        />
```

```
            <div dangerouslySetInnerHTML={{ __html: html
                }}></div>
            </div>
        </Layout>
    );
```

Use the `const` parameter that we defined in *Step 8* to render the image within a `GatsbyImage` component. Be sure to provide it with `alt` text to keep your image accessible – you could provide this via `frontmatter` as well if you wish.

10. Restart your development server and admire your hard work by navigating to your *about* page.

Now that we understand how we can source images within GraphCMS, let's turn our attention to how the same can be achieved in Prismic.

Sourcing images from Prismic

With a few simple changes to our configuration and queries, we can source images from Prismic utilizing `gatsby-plugin-image` and load them in on the site in the same way as local images:

1. Navigate to Prismic's website (`prismic.io`) and log in. Select your existing repository.

2. Navigate to the **CustomTypes** tab, click the **Create new** button, and select **Single Type**. Name your type **Profile** and submit it.

3. Using the **build-mode** sidebar (on the right), drag an image component into the type.

4. Name your field **photo**; the corresponding API ID should populate on its own. Click **OK** to confirm this. If you've done this correctly, your profile type should look as follows:

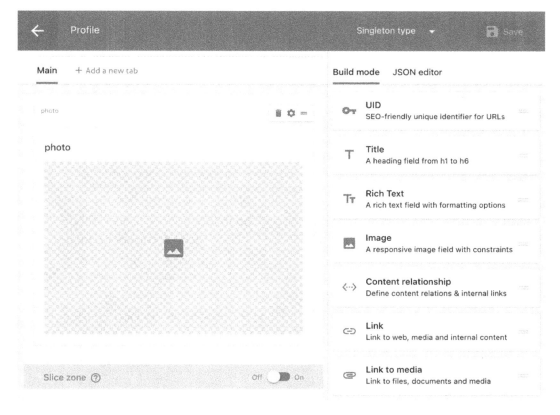

Figure 5.1 – Prismic profile type

5. Save the document.

6. Navigate to the JSON editor and copy its contents to a new file called `profile.json` within your `src/schemas` folder.

7. Navigate to the **Documents** tab and click the **Create new** button. If Prismic has not automatically opened your new type, select **Profile**. Using the interface, upload a new image into the document in the **photo** field.

8. Save and publish the new document. We now have everything set up in the CMS and can return to our Gatsby project.

9. Update your `gatsby-source-prismic` configuration within your `gatsby-config.js` file:

```
{
    resolve: "gatsby-source-prismic",
    options: {
```

```
        repositoryName: "elevating-gatsby",
        schemas: {
          icebreaker:
            require("./src/schemas/icebreaker.json"),
          profile:
            require("./src/schemas/profile.json"),
        },
        shouldDownloadFiles: () => true,
      },
    },
```

Add the new scheme to `schemas` that we added in *Step 6*. We will also add the `shouldDownloadFiles` option. This is a function that determines whether to download images. In our case, we always want it to download images so that we can use `gatsby-plugin-image`, and therefore set the function to always return `true`.

10. Add the `GatsbyImage` component and the `getImage` function as imports at the top of the `src/pages/about.js` file:

```
import { GatsbyImage, getImage } from "gatsby-plugin-image";
```

11. Modify your about page's `query` so that it includes the `prismicProfile` source:

```
export const query = graphql`
  {
    markdownRemark(frontmatter: { type: { eq: "bio" }
  }) {
      html
    }
    prismicProfile {
      data {
        photo {
          localFile {
            childImageSharp {
              gatsbyImageData(layout: FULL_WIDTH)
            }
          }
```

```
            }
          }
        }
      }
    ';
```

We utilize `gatsbyImageData` within our `query` and can make use of any of the configuration options that it supports. Here, we are specifying that the image should be full width.

12. Use the new data within the `data` prop to get the image within the component:

```
const {
    markdownRemark: { html },
    prismicProfile: {
      data: {
        photo: { localFile },
      },
    },
  } = data;
  const profileImage = getImage(localFile);
```

First, retrieve the `prismicProfile` data from the `data` prop and then use the `getImage` utility from `gatsby-plugin-image` to retrieve the image data that's required to render it and assign it to a `const` called `profileImage`.

13. Render your image within your `return` statement:

```
<GatsbyImage
        image={profileImage}
        alt="Your alt text"
        className="mx-auto max-w-sm"
      />
```

Use the `const` parameter defined in *Step 12* to render the image within a `GatsbyImage` component. Be sure to provide it with `alt` text to keep your image accessible – you could provide this via `frontmatter` as well if you wish.

14. Restart your development server and admire your hard work by navigating to your *about* page.

You should now feel comfortable using images and sourcing images via the Prismic CMS. While we have only looked at two CMSes, these concepts can be taken and applied to any headless CMS that supports images. Now, let's summarize what we have learned in this chapter.

Summary

In this chapter, we learned about images on the web, and how critical they are to our browsing experience. We learned about `gatsby-plugin-image` and implemented the two contained image components. We also learned in which circumstances to use which component.

At this stage, you should feel comfortable developing all manner of pages for your Gatsby site. In the next chapter, we will begin to look at how we can take our development site live. We will start this journey by looking at **search engine optimization**.

Part 2: Going Live

In this part, we will slowly move from working in development to getting our site ready to go live. By the end of this part, you should have your site deployed online.

In this part, we include the following chapters:

- *Chapter 6, Improving Your Site's Search Engine Optimization*
- *Chapter 7, Testing and Auditing Your Site*
- *Chapter 8, Web Analytics and Performance Monitoring*
- *Chapter 9, Deployment and Hosting*

6
Improving Your Site's Search Engine Optimization

In this chapter, you will learn about how **search engine optimization** (**SEO**) works, what search engines look for within your site pages, and how to improve your site's presence on the web. We will also dive into other uses of metadata to make visually enticing social share previews for your site. By the end of this chapter, you will have created a reusable SEO component to provide meta information on every page. We will also create a sitemap to make it easier for search engines to understand our site. Finally, we will also learn how to stop our site from appearing in search engines if you would rather not have it publicly visible.

In this chapter, we will cover the following topics:

- Introducing SEO
- Creating an SEO component
- Exploring meta previews
- Learning about XML sitemaps
- Hiding your site from search engines

Technical requirements

To complete this chapter, you will need to have completed *Chapter 5, Working with Images*.

The code for this chapter can be found at `https://github.com/ PacktPublishing/Elevating-React-Web-Development-with-Gatsby-4/ tree/main/Chapter06`.

Introducing SEO

SEO is the practice of improving the chances of search engines such as Google, Bing, and Yahoo, recommending your site's content to users as the best result for a given query or problem.

> **Important Note**
>
> Within this chapter, you will get an overview of what SEO is, why it's important, and how to implement pages with components that boost their SEO ranking. SEO is a vast subject and not something that can be covered in its entirety within this book. As such, you are encouraged to take what you learn in this chapter and build on it through your research.

Google will be the search engine that this chapter focuses on. Google has a 92% share of the search engine market worldwide. With all other search engines combined, taking less than a 10% share of the search engine market, there is no doubt that Google is dominating this space. Because of this, it is the logical search engine to gear this chapter toward.

If you want search engines to recommend your content, there is a trinity of tasks that need to be worked on in tandem:

- Ensuring your content can be discovered by search engine web crawlers.

- Showing search engines that you're a trustworthy source of information.

- Making your content user-friendly and inviting with a great UX, content hierarchy, and multimedia.

By investing time in implementing and improving these three things, the search engine will give you the most precious form of traffic – **organic traffic**. The best part? It's free! When Google shows your site as a part of a results page, you do not pay for its ranking or when it is clicked.

So, what's in it for the search engine? Ads and sponsors. Whenever search users search on Google, you will also be presented with results from sponsors that have been paid for, and occasionally personalized ads too. This is how search engines make their money and to keep their revenue stream consistent, they need you to keep coming back. To do that, they need to make sure that they bring you the best possible content for your search so that you use them again for your next search.

Now that we understand what search engines are looking for, let's learn about the important on-page signals that we can give to search engines to help them rank our site.

On-page search engine optimization

On-page signals are signals that a search engine can obtain from a site page. As your site pages are within your control, improving on-page signals is the easiest thing to get right and the easiest way to influence your site ranking. As a result, we will spend most of our time improving these signals in this chapter.

On-page signals can be split into two groups – technical and content signals.

Technical signals

Technical signals are those related to your site's code:

- **Speed**: Search engines want users to receive their results quickly, so pages that are fast receive a boost in ranking.

- **Mobile Responsiveness**: Most content on the web is consumed via mobile these days, so having a great mobile user experience is very important. Search engines are considering this more and more, with Google's index now being mobile-first.

- **Security**: Ensuring your website is secure improves your site's credibility. For example, **HTTPS** sites receive a boost over **HTTP** sites.

Content signals

Content signals are those related to the copy and links present on your site page:

- **Content Hierarchy**: The title, content headings, and page hierarchy are very important.

- **Page Content**: Google is always on the lookout for high-quality and accurate content that ultimately answers a user's query. Keep this in mind when you're creating pages and populating them. Your content needs to serve a real purpose for your site visitors.

- **Rich Content**: Google looks beyond the raw copy nowadays. The web is filled with multimedia content. Google is looking for content that contains images and videos instead of raw text. Multimedia content allows for better user interaction with your content and is therefore favored.

- **Recently Updated**: If the content within your pages has not been changed for a while, Google may treat its content as stale. Google is actively checking that the content of your page was created recently. By ensuring that your content is "fresh," Google can be sure your content is recent.

- **Outbound links**: By referencing content externally, this tells Google that the information is accurate as it is like content contained on multiple sites.

As you might be starting to realize by looking at these signals, you can sink vast amounts of time into these factors. It's up to you to decide how valuable search ranking and social media sharing are for your site, which will, in turn, determine how much time you should focus on implementing what is within this chapter.

Now that we understand what SEO is, let's turn our attention to how we can improve our on-page SEO ranking with an SEO component.

Creating an SEO component

Every site on the web has meta tags. Meta tags are snippets of text and image content that provide a summary of a web page. This data gets rendered in the browser whenever someone shares your site or when it appears within a search engine. Let's create an SEO component so that we can have rich previews that entice users to visit our site:

1. Install the necessary dependencies:

```
npm i react-helmet-async gatsby-plugin-react-helmet-async
```

   `react-helmet-async` is a dependency that manages all the changes that are made to your document head.

2. Include the `gatsby-plugin-react-helmet-async` plugin in your `gatsby-config.js` file:

```
module.exports = {
  // rest of config
  plugins: [
    `gatsby-plugin-react-helmet-async`,
  // other plugins
```

```
    ]
  }
```

This plugin updates your `gatsby-browser.js` file so that it wraps the root element in `HelmetProvider`, like this (*you do not need to do this step yourself*):

```
import React from "react";
import { HelmetProvider } from "react-helmet-async";


export const wrapRootElement = ({ element }) => {
  return (
    <HelmetProvider>
      {element}
    </HelmetProvider>
  );
};
```

We are now ready to edit the document's `head` within our React components and pages when required.

3. Create a new file inside `src/components` called `SEO.js`. This is the file in which we will create our SEO component.

4. Open the newly created file and add the following code:

```
import React from "react";
import { Helmet } from "react-helmet-async";
import { useStaticQuery, graphql } from "gatsby";
export default function SEO({ description, lang =
  "en", title }) {
  return (
    <Helmet
      htmlAttributes={{
        lang,
      }}
      title={title}
      titleTemplate={`%s · My Site`}
      meta={[
        {
          name: `description`,
          content: description,
```

```
        },
      ]}
    />
  );
}
```

Here, we are adding the language as an HTML attribute. We also added the `title` tag, a title template, and a `description` as metadata. The title template is useful if you want to have a consistent format. Let's imagine that the title we are passing in is `Home`. In this case, the template's final page title would be `Home · My Site`.

You are now ready to use your SEO component on your pages, so let's try it out! We will use the `src/pages/index.js` file as an example:

```
import React from "react";
import { Link, graphql } from "gatsby";
import { StaticImage } from "gatsby-plugin-image";
import Layout from "../components/layout/Layout";
import SEO from "../components/layout/SEO";

export default function Index({ data }) {
  const {
    site: {
      siteMetadata: { name, role },
    },
  } = data;

  return (
    <Layout>
      <SEO
        title="Home"
        description="The landing page of my website"
      />
      {/* REST OF FILE */}
    </Layout>
  );
}
```

Here, you can see the component embedded within our index page. We have added the `title` and `description` props to ensure these can populate the `<title>` and `<description>` tags contained within the SEO component. If you run `gatsby develop` at this point, you should see the title of the tab change to match your new title.

We can also provide `title` and `description` using data from our GraphQL data. In our blog page template file (`src/templates/blog-page.js`), we could use the blog post's `frontmatter` to populate the SEO component:

```
export default function BlogPage({ data }) {
  const {
    blogpost: {
      frontmatter: { date, tags, title, hero, desc },
      html,
    },
  } = data;
  return (
    <Layout>
      <SEO title={title} description={desc} />
      {/* Rest of render */}
    </Layout>
  )}
```

Here, we are passing `title` and `desc` from `frontmatter` of the Markdown post to the SEO component so that it can use these pieces of information to generate the tags.

Now that we have the basics set up, how do we enhance our site previews to make them more appealing within social media? Let's find out.

Exploring meta previews

If you've ever shared a website with a friend via Twitter, Slack, or any other instant messaging service, you probably saw a nice preview image, title, and description appear in a card to give the user insight into where you are sending them. This is achieved with meta tags.

We've already included a couple of these (`title` and `description` meta tags) within our search component, but here, we will implement two other common types – **OpenGraph** and Twitter metadata. We will then learn how to merge and validate these tags.

Open Graph metadata

Open Graph is an internet protocol that was originally designed and created by Facebook with a single purpose – to unify and standardize metadata within web pages to get better representations of the content of the page. The protocol does this by adding specific meta tags to your site header. These tags provide details about the content of your site pages. This could include information as basic as the page's title or maybe something more complex, such as how long a video on a page is. By populating the appropriate fields, we can create a bundled summary of what our site page looks like.

We can add Open Graph meta tags via our existing `SEO` component by adding them to the `meta` prop of the `Helmet` component:

```
<Helmet
    meta={[
        {
            property: `og:title`,
            content: title,
        },
        {
            property: `og:description`,
            content: description,
        },
        {
            property: `og:type`,
            content: `website`,
        },
        {
            property: `og:image`,
            content: `your-image-url.com`,
        }
    ]} />
```

Here, we are implementing Open Graph tags for the content's `title`, `description`, `type`, and `image`. As you can see, all open graph tags have the `og` prefix. These are just a few of the meta tags that are available via the protocol.

For a full list of all the available types, visit the Open Graph protocol website (`https://ogp.me`).

Twitter metadata

Like Facebook, Twitter also decided to create its own meta tags like Open Graph. All Twitter tags use the `twitter` prefix instead of `og`. One thing that separates Twitter tags from Open Graph's is that Twitter also has a tag for the content's display format on its platform. The first type is `summary`:

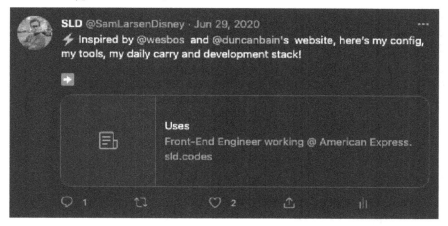

Figure 6.1 – Twitter summary card

`summary` shows a small summary preview of the site page. If you are looking for something larger with an image preview, you can use the `summary_large_image` type instead:

Figure 6.2 – Twitter summary card with a large image

As you can see, this type shows a larger image that is much more enticing to the user.

We can add Twitter meta tags via our existing `SEO` component by adding them to the `meta` prop of the `Helmet` component:

```
<Helmet
    meta={[
        {
            name: `twitter:card`,
            content: `summary_large_image`,
        },
        {
            name: `twitter:creator`,
            content: twitter,
        },
        {
            name: `twitter:title`,
            content: title,
        },
        {
            name: `twitter:description`,
            content: description,
        },
        {
            name: `twitter:image`,
            content: `your-image-url.com`,
        }
    ]} />
```

Though many of these tags are self-explanatory, it is worth calling out the `twitter:creator` tag. If you place your Twitter username as the content for this property, Twitter will be able to identify you as the creator of the site.

Now that we have implemented both Open Graph and Twitter meta tags, let's combine and merge the two.

Merging tags

You may have noticed that there is a little bit of duplication between the data we are providing via Twitter tags and Open Graph tags. For example, in both cases, we are providing a title (`twitter:title` and `og:title`). There is no harm in including these duplicates. Only a few bytes are added to your page by including this redundancy.

But if you would like to keep things clean, it is possible to reduce the number of tags. Twitter scrapes your site page – if it does not find the Twitter tag it is looking for, it will fall back to the Open Graph tags if they are present. This is great for duplicates such as the title and description, but it is still important to include those tags that are Twitter-specific, such as the card type.

Now, let's merge the Open Graph and Twitter tags we found in the previous two sections to create a subset that serves both formats without redundancy:

```
<Helmet
    meta={[
        // Twitter MetaData
        {
            name: `twitter:card`,
            content: `summary_large_image`,
        },
        {
            name: `twitter:creator`,
            content: twitter,
        },
        // Open Graph MetaData
        {
            property: `og:title`,
            content: title,
        },
        {
            property: `og:description`,
            content: description,
        },
        {
            property: `og:type`,
            content: `website`,
        },
```

```
      {
        property: `og:image`,
        content: `your-image-url.com`,
      },
    ]}
  />
```

Here, we can see that we have completely omitted the Twitter tags for `title`, `description`, and `image` as they will fall back to the Open Graph tags. We have, however, retained the Twitter tags for `creator` and `card` as they are not available via Open Graph.

Now that we understand how to make our site look great when we share it, what about when it is shared for us by a search engine? How do we highlight the information we want it to care about?

Validating tags

Regardless of whether you have implemented Open Graph tags, Twitter tags, or both, you will want to ensure that your tags are working correctly before sharing your site pages online. Both Facebook and Twitter have created applications to preview how links that are shared on their platforms will be displayed:

- Twitter Card Validator: `https://cards-dev.twitter.com/validator`
- Facebook Sharing Debugger: `https://developers.facebook.com/tools/debug`

These tools perform a very similar function – they scrape the web page that's entered for any relevant meta tags that you have defined. Then, they display what a preview of the site would look like on their platform using these tags. There are also third-party services that will validate for both of these platforms at the same time, such as MetaTags.io (`https://metatags.io`).

> **Tip**
> The validators mentioned here only work with sites that are hosted on public servers. You will not be able to test your meta tags using these without deploying your site first. You will learn more about how to deploy your site in *Chapter 9, Deployment and Hosting*.

At this point, you should feel comfortable with creating and testing meta tags. Now, let's turn our attention to how we can make our site easier for web crawlers to interpret.

Learning about XML sitemaps

A sitemap is a special file that provides information about the web pages and files on your site, as well as their relationships. Creating this file allows web crawlers to gather information about your site without having to crawl your site manually. It helps us highlight to search engines which pages we specifically want them to look at. Let's create a sitemap for our site:

1. Install the `gatsby-plugin-sitemap` dependency:

    ```
    npm install gatsby-plugin-sitemap
    ```

2. Update your `gatsby-config.js` file:

    ```
    module.exports = {
      siteMetadata: {
        siteUrl: `https://your.website.com`,
      },
      plugins: [
        `gatsby-plugin-sitemap`,
      // other plugins
      ]
    }
    ```

    By including this plugin, Gatsby will automatically create a sitemap when building the site. It's important to remember that this plugin only generates output when running in production. When you're using `gatsby develop`, you will not see your sitemap file being created. Only when the `gatsby build` command is running will the sitemap file be created.

Now that we have followed those steps, let's verify our implementation. Run `gatsby build && gatsby serve` and navigate to `/sitemap/sitemap-index.xml` on your site. This page should show you some **XML** data that looks like this:

```
<sitemapindex
  xmlns="http://www.sitemaps.org/schemas/sitemap/0.9">
<sitemap>
<loc>https://your.website.com/sitemap/sitemap-0.xml</loc>
</sitemap>
</sitemapindex>
```

This page is the index of your sitemap and tells search engines where to find your site data. In your case, you will probably see a single entry, similar to the one shown in the preceding code block. If you follow the path in the `<loc>` tag (`/sitemap/sitemap-0.xml`), you will find something like this:

```
<urlset>
<url>
<loc>https://your.website.com/blog</loc>
<changefreq>daily</changefreq>
<priority>0.7</priority>
</url>
<url>
<loc>  Post/</loc>
<changefreq>daily</changefreq>
<priority>0.7</priority>
</url>
...
</urlset>
```

In this list, you should see an entry for each page on the site, with a `changefreq` and a `priority`. If the site pages appear in this list, then Google can find information about your pages without having to manually crawl your site. Congratulations!

> **Important Note**
> Your site's 404 page and its development variants are always excluded from the sitemap, so you don't have to worry about filtering these pages out.

Google has built a great tool for validating your sitemap, as well as other search analytics – Google Search Console. You can use it to check your site's indexing status and optimize the visibility of your site based on which queries are driving traffic to your website. You can try it out by visiting `https://search.google.com/search-console/about`.

We now understand how to make it easy for Google to find and display our site, but what if we want to do the opposite and stop our site from appearing in search engines? We'll look at this in the next section.

Hiding your site from search engines

To prevent your page from appearing within Google and other search engines, you must update the <head> property of the page so that it includes the following meta tag:

```
<meta name="robots" content="noindex">
```

By including a noindex meta tag, crawlers that crawl that page and see the tag will drop the page from their search results. This happens regardless of whether the site is being linked to any other site on the internet.

Much like our SEO component, we could make this addition to a component so that we can reuse it across the pages when needed:

1. Create a new file in src/components/layout called NoRobots.js.
2. Open the newly created file and add the following code to it:

    ```
    import React from "react";
    import { Helmet } from "react-helmet-async";

    export default function NoRobots() {
      return (
        <Helmet
          meta={[
            {
              name: `robots`,
              content: "noindex",
            },
          ]}
        />
      );
    }
    ```

This component adds the noindex metadata to head when it's included on any page. Using this component in this way gives us the flexibility to make a few pages hidden while still allowing the rest to be indexed.

To hide static resources such as images within your site from search engines, we need to include a `robots.txt` file. This file is used by search engine crawlers to determine which parts of your site it can access. There is a plugin called `gatsby-plugin-robots-txt` that has been set up to make creating this file painless. Let's implement this plugin now:

1. Install the `gatsby-plugin-robots-txt` dependency:

    ```
    npm install gatsby-plugin-robots-txt
    ```

2. Update your `gatsby-config.js` file with the following code:

    ```
    module.exports = {
      siteMetadata: {
        siteUrl: `https://your.website.com`,
      },
      plugins: [
        resolve: 'gatsby-plugin-robots-txt',
          options: {
            host: ' https://your.website.com,
            sitemap: https://your.website.com
            /sitemap.xml',
            policy: [{ userAgent: '*', disallow: '/static'
            }]
          }
    // other plugins
      ]
    }
    ```

    Within this plugin's configuration, we have specified that all crawler user agents can access the whole site except for the `/static` folder. The `/static` folder is where site images are stored. Adding this `disallow` policy will stop your images from appearing in a Google image search. They will only be found within your site pages.

We now have a clear understanding of how to omit both pages and assets from search engines.

Summary

In this chapter, we learned what SEO is and what signals search engines use to identify quality content. We created an SEO component that we can use to add meta information to our site pages. Then, we enhanced this with Open Graph and Twitter meta tags for better site previews on social media platforms. We also implemented a sitemap to help search engines index our site effectively. Finally, we learned how to hide our site from search engines.

In the next chapter, we will learn how to test and audit our site. We will also explore how to audit our site page's SEO. By learning how to audit our site pages, we can identify ways to improve the speed of our pages too, which will also boost our SEO ranking.

7
Testing and Auditing Your Site

In this chapter, we will learn about what unit testing is, why it's useful, and how to start unit testing your Gatsby site. We will then learn how we can use Git hooks to trigger your unit tests and other commands when running common Git commands. Following this, we will investigate how we can measure core web vitals to understand how well our Gatsby site's page experience is performing, both in lab and field environments. By the end of this chapter, you should feel comfortable that you can analyze how well a Gatsby site is working locally by using unit tests and looking at web vitals when it is out there on the web.

In this chapter, we will cover the following topics:

- Exploring unit testing
- Adding Git hooks for tests
- Auditing core web vitals

Technical requirements

To complete this chapter, you will need to have completed *Chapter 6, Improving Your Site's Search Engine Optimization*. You will also need Google Chrome installed.

The code for this chapter can be found at `https://github.com/ PacktPublishing/Elevating-React-Web-Development-with-Gatsby-4/ tree/main/Chapter07`.

Exploring unit testing

Unit testing is a way of testing the smallest piece of code that you have logically defined within your application. During unit testing, we isolate a small part of the code and verify that it is behaving as intended independently from the rest of the code base. We instantiate this piece of code, invoke it, and then observe its behavior. If the observed behavior matches what we expected, then we know that our code is doing what it should be. By setting up a multitude of these tests, we can have a better understanding of where something has broken when we edit large parts of the code base.

Within React and Gatsby, there are multiple different ways in which you can set up unit tests. Here, we will focus on one of the most popular combinations – **Jest** and **React Testing Library**. Let's create a structure within our repository that will allow us to test our site using these tools:

1. Install the necessary dependencies:

```
npm install -D jest babel-jest @testing-library/jest-
dom @testing-library/react babel-preset-gatsby
identity-obj-proxy
```

2. Create a `jest.config.js` file:

```
module.exports = {
  transform: {
    "^.+\\.jsx?$": '<rootDir>/jest-preprocess.js',
  },
  moduleNameMapper: {
    ".+\\.(css|styl|less|sass|scss)$": 'identity-obj-
      proxy',
    ".+\\.(jpg|jpeg|png|gif|eot|otf|webp
      |svg|ttf|woff|woff2|mp4|webm|wav|mp3|m4a|aac
      |oga)$": '<rootDir>/__mocks__/file-mock.js',
```

```
  },
    testPathIgnorePatterns: ['node_modules', '.cache'],
    transformIgnorePatterns:
      ['node_modules/(?!(gatsby)/)'],
      testEnvironment: "jsdom",
      globals: {__PATH_PREFIX__: '',
    },
    setupFiles: ['<rootDir>/loadershim.js'],
    setupFilesAfterEnv: ['<rootDir>/jest.setup.js']
  };
```

Both Gatsby and Jest use Babel under the hood. However, unlike Gatsby, Jest does not handle its own Babel configuration. We use the `jest.config.js` file to manually set up Jest with Babel, as well as configure our tests.

Let's break down the contents of this file so that we understand what each part is doing:

a. `transform`: This tells Jest that all the files that end in `.js` or `.jsx` need to be handled with a `jest-preprocess.js` file, which we will create in the next step.

b. `moduleNameMapper`: When testing, it is uncommon to test static assets such as images. As such, Jest does not care for them. But it is still important that it knows how to handle them as they may be embedded in your code. Here, we are giving Jest a mock for handling stylesheets that uses the `identity-obj-proxy` package, which we installed in the first step, and another mock that handles common image, video, and sound files. We will create this second mock later in this section.

c. `testPathIgnorePatterns`: This tells Jest to ignore any tests found within `node_modules` as we do not want to bring in tests that have been found within our packages and the `.cache` directory.

d. `transformIgnorePatterns`: Here, we tell Jest to ignore Gatsby when it is transforming code, as Gatsby includes untranspiled ES6 code.

e. `globals`: This is where we define a global variable called `__PATH_PREFIX__` that Gatsby uses behind the scenes. We need to define it here too as some Gatsby components will break without it being present.

f. `setupFiles`: Here, we list the configuration files that we would like to use to configure the testing environment. It is run once per test. Here, we tell it to run `loadershim.js`, which we will create later in this section.

g. `setupFilesAfterEnv`: Here, we specify the configuration files we would like to use to set up our tests. Crucially, these files run after the testing environment has been set up.

3. Create a `jest-preprocess.js` file within your root directory:

```
const babelOptions = {
  presets: ["babel-preset-gatsby"],
};

module.exports = require("babel-
  jest").default.createTransformer(babelOptions);
```

This is where we define our Babel configuration. As we are working with Gatsby, we are using the `babel-preset-gatsby` preset. You can expand this preset list as necessary.

4. Create a `loadershim.js` file within your root directory:

```
global.___loader = {
    enqueue: jest.fn(),
  }
```

We use this file to mock a global `loader.enqueue` function using a Jest mock function.

5. Create a new folder in your root directory called `__mocks__`.

6. Create a `file-mock.js` file within the `__mocks__` folder:

```
module.exports = "test-file-mock"
```

As we mentioned in *Step 2*, this file mocks out static asset file types.

7. Create a `gatsby.js` file within the `__mocks__` folder:

```
const React = require("react")
const gatsby = jest.requireActual("gatsby")
module.exports = {
  ...gatsby,
  graphql: jest.fn(),
```

```
Link: jest.fn().mockImplementation(
  ({
    activeClassName,
    activeStyle,
    getProps,
    innerRef,
    partiallyActive,
    ref,
    replace,
    to,
    ...rest
  }) =>
    React.createElement("a", {
      ...rest,
      href: to,
    })
),
StaticQuery: () => React.createElement("div", {
  id: "StaticQuery",
}),
useStaticQuery: jest.fn(),
}
```

Here, we are mocking out any components or functions that we are using from the gatsby package. We strip the props from the Link component and return an <a/> tag instead. We return a div in place of the StaticQuery component. Finally, we also mock out the useStaticQuery function with a Jest mock function.

8. Create a gatsby-plugin-image.js file within the __mocks__ folder:

```
const React = require("react")
const gatsbyPluginImage = jest.requireActual("gatsby-
  plugin-image")

module.exports = {
  ...gatsbyPluginImage,
  StaticImage: () => React.createElement("div", {
```

```
        id: "StaticImage",
    }),
}
```

Here, we are mocking out any components or functions that we are using from the gatsby-plugin-image package. We return a div in place of the StaticImage component.

9. Create a jest.setup.js file within your root directory:

```
require('@testing-library/jest-dom/extend-expect');
const { useStaticQuery } = require("gatsby");

beforeAll(() => {
  useStaticQuery.mockReturnValue({
    site: {
      siteMetadata: {
        siteUrl: "test.url.com",
        social: { twitter: "@slarsendisney" },
      },
    },
  });
});
```

Before each test, we need to mock a return value for useStaticQuery. Any page components that make use of the SEO component will fail unless they can retrieve this data from the function.

10. Create a test-utils.js file within your src directory:

```
import React from 'react'
import {render} from '@testing-library/react'
import { HelmetProvider } from "react-helmet-async";

const Wrapper = ({children}) => {
  return (
    <HelmetProvider>
        {children}
    </HelmetProvider>
  )
```

```
    }

    const customRender = (ui, options) =>
        render(ui, {wrapper: Wrapper, ...options})

    export {customRender as render}
```

This file is not required but is helpful. Much of your application might make use of a provider, which we would normally wrap our root element in, within `gatsby-browser.js`. We can't do that in Jest. So, instead of defining `wrapper` in every test, it is preferable to create a custom `render` function that wraps any content in the required providers. We then call this `render` instead of the one that's exported from `@testing-library/react` when required.

11. Create a test script within your `package.json` file:

```
"scripts": {
    "build": "gatsby build",
    "develop": "gatsby develop",
    "start": "npm run develop",
    "serve": "gatsby serve",
    "clean": "gatsby clean",
    "test": "jest"
},
```

Now, when using the `npm run test` command, it will start Jest and begin testing.

We now have everything in place to start testing! This book does not have space for a full guide on unit testing. However, let's create a few example tests for a few different component types, such as simple components, SEO components, and our Gatsby page components.

Testing simple components

Testing simple components can be done in the same way you would do so in any standard react project. Let's take a look at how we would test our header component, as an example.

Create a `Header.test.js` file next to your header component using the following code:

```
import React from "react";
import {render, screen} from '@testing-library/react'
import '@testing-library/jest-dom'
```

```
import Header from "./Header";

test("Renders header", async () => {
  render(<Header />);


  expect(screen.getByText('Site Header'))
});
```

Here, we are rendering our `Header` component to the screen. We are then testing that the screen contains some text stating `Site Header` to ensure that the `Header` component is rendered. We do this by using the `screen.getByText` function.

Now that we understand how to test simple components, let's look at a more complex example – your site's SEO component.

Testing the SEO component

A common component among Gatsby pages that is important to test is the SEO component. It is important to ensure that the meta tags we are adding using the component are being correctly applied to the document's head so that we know that when that page is shared, it will have the rich previews that we set up in *Chapter 6, Improving Your Site's Search Engine Optimization*. Let's look at how we could test this.

Create a `SEO.test.js` file next to your SEO component using the following code:

```
import React from "react";
import { render } from "@testing-library/react";
import "@testing-library/jest-dom";
import { HelmetProvider } from "react-helmet-async";
import SEO from "../SEO";

HelmetProvider.canUseDOM = false;

test("Correctly Adds Meta Tags to Header", async () => {
  const mockTitle = "Elevating React with Gatsby";
  const mockDescription = "A starter blog demonstrating
    what Gatsby can do.";
  const context = {};
  render(
    <HelmetProvider context={context}>
```

```
  <SEO title={mockTitle} description={mockDescription} />
  </HelmetProvider>
);

  const head = context.helmet;
  expect(head.meta.toString()).toMatchSnapshot();
});
```

First, we inform `react-helmet-async` from `HelmetProvider` that it cannot use the **Document Object Model (DOM)** as this is only available in the browser. This allows it to emulate how it behaves when your site is being built. Within the test itself, we first create a mock title and description. Then, we pass this to the `SEO` component. After rendering, we check that the context's helmet object contains meta, and if it does, we make sure it matches the snapshot.

Now, let's understand how we would test whole site pages.

Testing Gatsby page components

If you ever want to test pages, you can make use of the custom `render` function we set up in *Step 10* of the *Exploring unit testing* section. Let's take a look at how we would test our site's index page as an example.

Unlike our component tests, it is best to avoid placing your page tests in the same directory as the page files. This is because Gatsby will automatically try and create pages for every exported React component in the `pages` directory. Instead, create a folder alongside the `pages` directory called `pages-lib` that's specifically for Gatsby page tests.

Create an `index.test.js` file in the `pages-lib` directory using the following code:

```
import React from "react";
import { screen } from "@testing-library/react";
import {render} from "../../test-utils"
import "@testing-library/jest-dom";
import Index from "../pages/index";

test("Renders Index Page with correct name", async () => {
  const data = {
    site: {
      siteMetadata: { name: "My Name", role: "My Role" },
    },
```

```
  };
  render(<Index data={data} />);

  expect(screen.getByText(data.site.siteMetadata.name));
});
```

In this instance, we are making use of the custom render function that we set up in the test-utils.js file. This is because page components typically also contain an SEO component, which uses the Helmet component, and, as such, needs to be wrapped in HelmetProvider. It's also important to pass any data to the data prop that the page would normally retrieve via GraphQL, as GraphQL queries on the page will not run.

Now that we understand how to write tests, let's understand how we can trigger them with Git hooks.

Adding Git hooks for tests

A **Git hook** is a method that fires on common Git commands. We can use this method to invoke custom scripts when we commit or push our code. It is common practice to use these hooks to run checks against your repository to ensure that the code that is being added does not break application functionality. One valuable check we could add is to run our applications unit tests before we push our code, and if they fail, we can stop the push. By implementing this feature, it's unlikely that the code being pushed will break any functionality we test for.

Let's implement this functionality now by creating a Git hook that is triggered by a git push. This will ensure our unit tests pass before allowing the push command to run. We will be using the husky package to do this as it is easy to set up and maintain:

1. Install the necessary dependencies:

    ```
    npm install husky --save-dev
    ```

2. Create a postinstall script in your package.json file with the following command:

    ```
    npm set-script postinstall "husky install"
    ```

    This command will add a new script to our package.json file called postinstall that causes husky to be installed.

3. Run this new script:

    ```
    npm run postinstall
    ```

As we are setting this up for the first time, we will need to trigger the `husky` install manually by running the `postinstall` script via the command line. Every subsequent developer will never need to run this manually.

4. Add a hook:

```
npx husky add .husky/pre-push "npm run test"
```

This adds a `pre-push` hook that runs our npm `test` script. After running this command, every subsequent push will run the `test` script and only push on success.

> **Important Note**
> Running tests on push is not always the best test. We may have uncommitted code locally that is causing the tests to pass, which are not included in the push. This can cause the same tests to fail in **continuous integration/continuous deployment (CI/CD)** environments.

Now that we understand how to trigger unit tests with Git hooks, let's turn our attention to a different kind of test – auditing core web vitals.

Auditing core web vitals

Web vitals (`https://web.dev/vitals`) are an initiative by Google to provide unified guidance for quality signals that are essential to delivering a great user experience on the web. These directly tie into the signals discussed in *Chapter 6, Improving Your Site's Search Engine Optimization*.

The core web vitals are a small group of Google's web vitals that focus on three pillars – how fast the page loads, how soon you can interact with the page, and how stable the page is while it is loading and while the user is interacting with it. These three pillars are encompassed in the following three metrics:

- **Largest Contentful Paint**: A representation of load time. It is the measure of the time the browser takes to make the majority of a page's content visible from the moment you start navigating to it. This is the moment at which a user perceives the site to have finished loading.

- **First Input Delay**: Measures the response time to interact. First input delay is the time the browser takes from navigation to a point where you can interact with any element on the page, such as a form or button.

- **Cumulative Layout Shift**: A measurement of how stable the page is while it loads. The less your elements shift around the page while the page loads, the higher your score will be.

Now that we understand these core web vitals, how do we measure them? There are two different methods we can use to retrieve these metrics. They are as follows:

- **Lab Test Data**: Data generated on demand by you for testing. This is less accurate as it is based on approximations of user data. But while developing, it is often incredibly useful as we can use it to develop our site iteratively, without ever having to deploy it.

- **Field Data**: Data collected from users when viewing your site. This is the most accurate source of data as it directly corresponds to how your users are perceiving your site.

Let's look at how we can retrieve lab test data using the lighthouse tool and field data using the `web-vitals` package.

Using Chrome's lighthouse tool

The lighthouse tool will analyze your website for performance, accessibility, search engine optimization, and progressive web app features. Not only does it give you a score in each of these categories, but it will also tell you how to improve your site to increase these scores. The best part? It's built into Google Chrome – no other downloads or tooling installation is required.

Now, let's generate a lighthouse report for our site using the tool:

1. Build your site using the `gatsby build` command. As we saw in *Chapter 1, An Overview of Gatsby.js for the Uninitiated*, this creates a production build of your website. It is *vitally* important that we audit a production build of the site instead of a development build, since the development tools that Gatsby adds into the build drastically increase your site's bundle.

2. Serve your build using the `gatsby serve` command. With the default settings, your site should be live on `http://localhost:9000/`.

3. Open Google Chrome in incognito mode and navigate to `http://localhost:9000/`. You should see the index page of your site. By loading this page in incognito mode, we ensure that no Chrome extensions you have installed interfere with the test.

4. Right-click anywhere on the page and click **Inspect**. This will bring up **Developer Tools** on the right-hand side of the window.

5. Click the chevron in the center of the top bar and select **Lighthouse**:

Figure 7.1 – Lighthouse location Within Developer Tools

6. This will present you with the lighthouse report generator window, which looks like this:

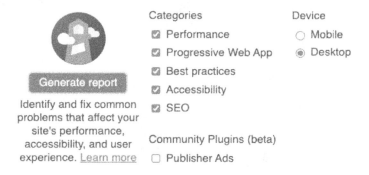

Figure 7.2 – Lighthouse report generator

Select the categories you would like to audit, all of which will be switched on by default. All of these categories are important, and it is advised to keep them all on unless you are specifically trying to improve a single metric and want the report to be generated more quickly.

You must also select a **Device** type. By default, this setting is set to **Mobile**. Lighthouse will try to emulate a mobile device attempting to access the page, which includes using a smaller viewport and throttling the network connection. Running multiple reports – one for each device type – is a great idea as it ensures that your site has a great experience on every device. Note that for SEO purposes, Google uses mobile metrics in their site rankings.

7. Clicking **Generate report** will start lighthouse. You may see the page flash a few times during this process. This is nothing to worry about. Congratulations – you've just run your first lighthouse report!

Once lighthouse has finished running, you will see that the report generator window has been replaced with a report that contains a section for each category. Let's take a look at the **Performance** category:

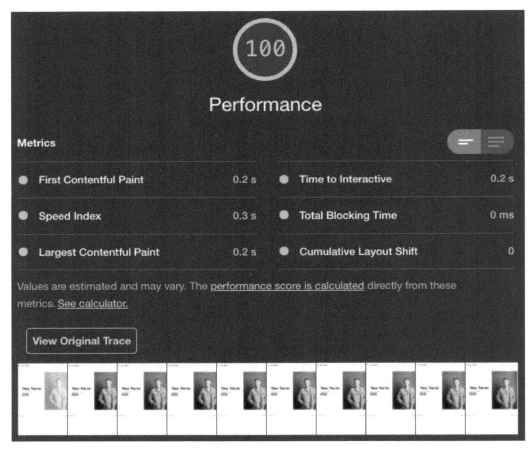

Figure 7.3 – Lighthouse Performance report

You may notice some familiar items within these metrics. Lighthouse has audited each of the three core web vitals as part of its audit. Each metric will be color-coded to give you an indication of where you need to focus your efforts. Green means good, orange means that it needs improvement, while red means that the score for this metric is considered poor. In cases where your scores are not optimal, lighthouse will propose changes that you can make to your site to improve the score. Let's look at an example:

Accessibility

These checks highlight opportunities to improve the accessibility of your
web app. Only a subset of accessibility issues can be automatically
detected so manual testing is also encouraged.

Names and labels — These are opportunities to improve the semantics of the controls in your
application. This may enhance the experience for users of assistive technology, like a screen
reader.

⚠ Buttons do not have an accessible name   ∧

When a button doesn't have an accessible name, screen readers announce it as "button",
making it unusable for users who rely on screen readers. Learn more.

Failing Elements

`button.theme-standard`

Figure 7.4 – Lighthouse Accessibility report with suggested improvements

In the preceding screenshot, we can see that our buttons are not currently accessible as
they do not have accessible names. Hovering over the failing element will highlight it
within the site so that we can rectify it quickly.

Now that we understand how to retrieve lab test data, let's investigate how we can retrieve
field data using the web-vitals JavaScript package.

Using the web-vitals JavaScript package

The web-vitals package is a 1 KB package that's developed by the Google Chrome
team. This package monitors web vitals, including core web vitals on real users as they
visit your site. It aims to measure them in a way that is incredibly similar to other Google
reporting tools.

> **Important Note**
> The `web-vitals` package makes use of browser APIs that are not supported in all browsers. The package only guarantees complete support in Google Chrome. If you are gathering metrics using this tool, please consider that the results will only be retrievable on supported browsers. If you are collating these metrics, it is important to remember that they do not necessarily represent all your site visitors.

To understand how to use `web-vitals` within our application, let's create a rudimentary example where we simply log the vitals when the user navigates to our site:

1. Install the `web-vitals` package:

    ```
    npm install web-vitals
    ```

2. Create a function that utilizes the `web-vitals` package:

    ```
    import { getCLS, getFID, getLCP } from "web-vitals";

    export default async function webVitals() {
      try {
        getFID((metric) => console.log(metric));
        getLCP((metric) => console.log(metric));
        getCLS((metric) => console.log(metric));
      } catch (err) {
        console.log(err);
      }
    }
    ```

    In this example, we retrieve the metrics and then simply log them to the console. It is important to wrap them in a `try catch` block to avoid crashing the page when the APIs are not supported. This also allows you to handle the error accordingly.

3. Use the following code within your `gatsby-browser.js` file:

    ```
    import "./src/styles/global.css"
    import webVitals from "./src/utils/web-vitals"

    webVitals()
    ```

By calling the function within this file, it will run once when the user initially navigates to our site from an external source, but not on every page navigation within the site.

Starting your development server and navigating to your site via Google Chrome, you should see the metrics logged in the console. In this example, we are simply displaying them, but we could be sending these to our analytics platform. We will look at this in more detail in *Chapter 8, Web Analytics and Performance Monitoring*.

We now have a good understanding of how to measure web vitals both in the field and during development.

Summary

In this chapter, we learned about unit testing – what it is and why it is important. Then, we integrated unit testing into our Gatsby site. We also looked at a few different recipes for unit tests that we can use to test different types of react components. We then learned about Git hooks and implemented a Git hook that runs unit tests using husky. Finally, we investigated core web vitals. We used web vitals to test our page experience both locally using lighthouse, and in the field using the web-vitals package. Using what you've learned, you should now feel that you can test a site locally, as well as audit its performance, accessibility, and SEO once it is out on the web.

In the next chapter, we will discover how we can add analytics to our site, including how we can track web-vitals field data.

8
Web Analytics and Performance Monitoring

In this chapter, we will investigate ways in which we can monitor the behavior of our application and the users who visit it. We will learn how page analytics can be a valuable tool in helping create better user experiences. We will learn how to use two different tools to gather these page insights, depending on what we require. We will learn what is required of us legally when using these tools. We will also implement plugins that allow us to debug errors that our users encounter when using our site.

By the end of this chapter, you should feel confident that you can gather different types of analytics and use them to inform yourself (or the site owner) about how code changes have affected the user experience.

In this chapter, we will cover the following topics:

- Introducing website analytics
- Implementing page analytics
- Monitoring the performance of your site

Technical requirements

To complete this chapter, you will need to have completed *Chapter 7, Testing and Auditing Your Site*.

The code for this chapter can be found at `https://github.com/ PacktPublishing/Elevating-React-Web-Development-with-Gatsby-4/ tree/main/Chapter08`.

Introducing website analytics

Website analytics is the act of collecting, aggregating, analyzing, and reporting a website's data. Let's break website analytics down into two categories:

- **Page analytics**: Analytics we gather about how users interact with our website. This could be page views, click rates, or bounce rates, for example.

- **Performance Monitoring**: Analytics we gather on how our code performs for our users. This is primarily used for logging errors in our JavaScript that our users encounter.

Regardless of the category, they all work in a similar way. First, an inserted `script` tag loads a small amount of JavaScript into the page. This code is run in the web browser of anyone visiting the site. In most cases, the code drops a small text file with small pieces of data known as cookies onto the users' browsers. This data is used to identify the user session. This is sent back to the analytics tool, along with request information, to identify the user and the event that is being tracked.

You've collected all this data – now what? By looking at aggregated users' data, we can gain insights into how our users are behaving. This information is the most powerful ally you can have when you're trying to improve your user experience. You can use these reports to identify trends in the kind of content your users like or pages where users most often leave your application, for example.

Now, let's take a moment to talk about privacy concerns when gathering user data.

Privacy

Regardless of what data you intend on collecting, it is always important to consider the privacy of your users. A published and publicly accessible privacy policy is a legal requirement if you intend to store, transfer, or handle a user's personal information. It is also the case that the popular analytics providers, including *Google Analytics*, specify in their terms of service that if you are using their service, you must publish a privacy policy. If you do not, you are in breach of your contract with them and are using the tool illegally.

In addition to a privacy policy, it is also a good idea to have a cookies policy, if your website has visitors from Europe. As of 2018, the European Union state that you are required to get "clear, informed consent" from your users to use cookies. This normally takes the form of a banner that you display to users on their first visit to your site.

Now that we understand what website analytics is, let's turn our attention to how we can implement the first type of them – page analytics.

Implementing page analytics

There is a multitude of tools you can use to perform page analytics. In this book, we are going to look at the following two:

- Google Analytics
- Fathom Analytics

Google Analytics is the world's leader in page analytics. More than half of all websites on the web use this tool. One of the reasons for its popularity is its age as it has been around since 2005. When it was launched, there wasn't much competition in the analytics space. Another reason for its popularity is that it's free. It's important to remember that if you are using a free tool or site, it is often the case that your data is the product. If you are concerned about your privacy and that of your site visitors, then perhaps Fathom Analytics is a better choice.

Unlike Google, **Fathom Analytics** does not track personal data. For example, when a page view is logged, it only tells you it was visited, but not by who specifically. Fathom's script is cookie-free, which means that you do not need a cookie policy or cookie consent banner. As Fathom is determined to collect as little personal information as possible, your privacy policy can also be considerably shorter.

> **Important Note**
>
> Only implement one of the page analytics tools mentioned in this section. Having multiple scripts that accomplish the same thing is only going to make your page heavier.

We will discuss how to implement both page analytics tools in the following sections. Let's start with Google Analytics.

Adding Google Analytics

To start tracking data within our Gatsby site, we will need to obtain a measurement ID from Google Analytics. Let's do this by following these steps:

1. Navigate to `https://analytics.google.com/analytics/web/?authuser=0#/provision/create` from your browser.

2. Give your account an **Account name** and enter your **Account Data Sharing Settings** preferences:

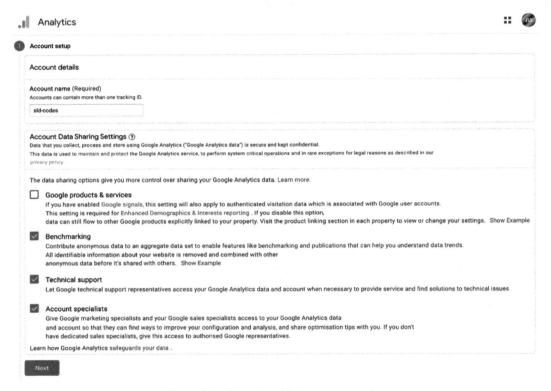

Figure 8.1 – Google Analytics account setup

The name is specific to a project, so call it something relevant. Pay attention to the data sharing options – only ever share with Google what you're comfortable sharing.

3. Set up a property by entering a **Property name**, **Reporting time zone** (region), and **Currency**:

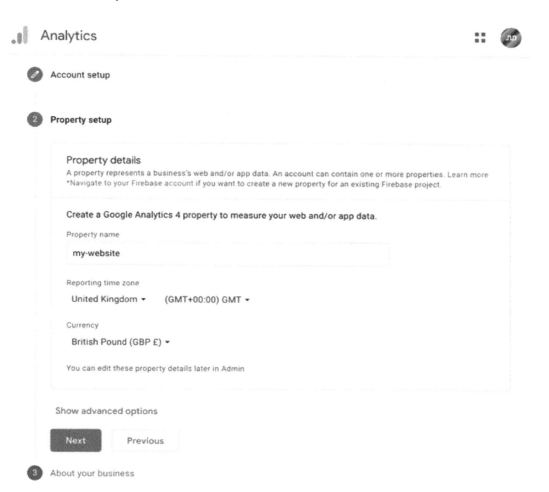

Figure 8.2 – Google Analytics property setup

This property is specific to your website or app. In our case, this will be used to reference analytics from our Gatsby site, so a name such as **personal-website** or **my-website** would be appropriate.

4. Fill out the business information that Google requires and submit the form. You will then be presented with the following screen:

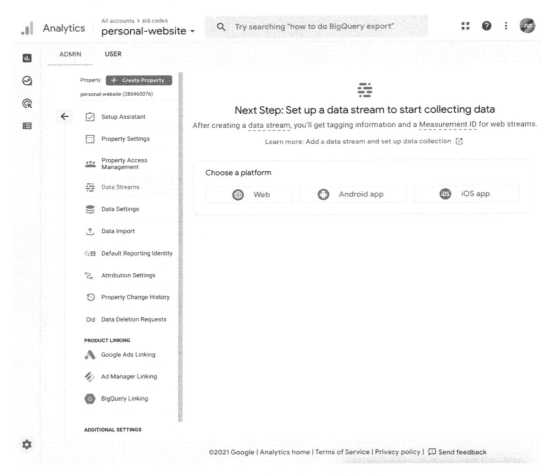

Figure 8.3 – Google Analytics dashboard

This is your first look at the Google Analytics dashboard. Before we can start leveraging its power, we will need to set up our first data stream.

5. Select **Web** under the **Choose a platform** heading. This will open the following screen:

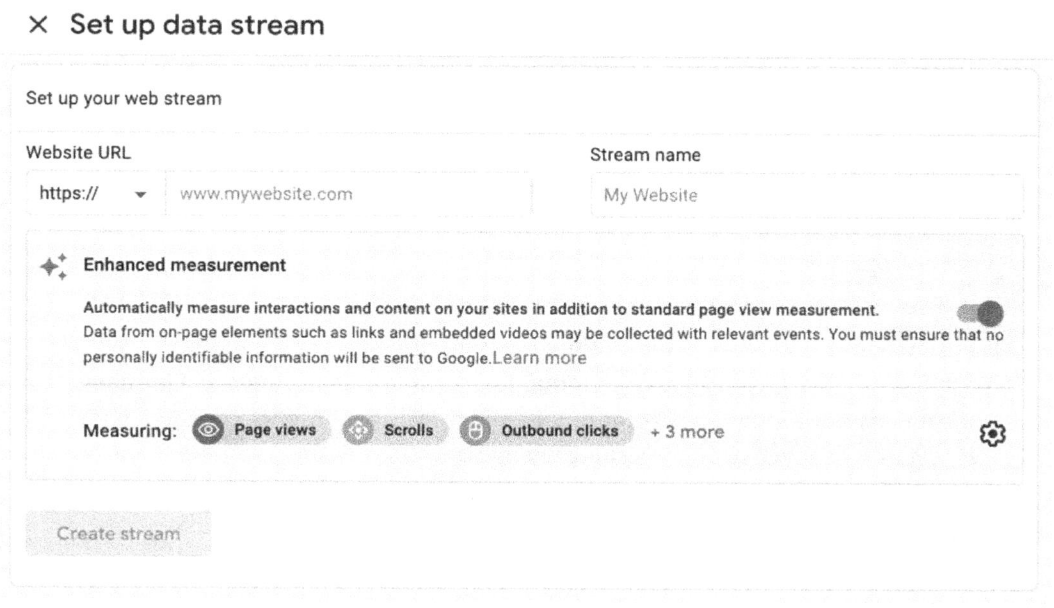

✕ Set up data stream

Set up your web stream

Website URL

Stream name

https:// ▾   www.mywebsite.com

My Website

✦ **Enhanced measurement**

Automatically measure interactions and content on your sites in addition to standard page view measurement.
Data from on-page elements such as links and embedded videos may be collected with relevant events. You must ensure that no
personally identifiable information will be sent to Google. Learn more

Measuring: ◉ **Page views**   ⚙ **Scrolls**   ⊕ **Outbound clicks**   + 3 more   ⚙

Create stream

Figure 8.4 – Google Analytics web stream setup

Enter your website address under **URL** and **name** your web stream. Finally, submit the form by clicking **Create stream**. This will bring up the details of our newly created stream.

6. Make a note of the stream's measurement ID.

Now that we have obtained the measurement ID, let's turn our attention to our Gatsby site repository and use it to start gathering site statistics:

1. Install the necessary dependencies:

```
npm install gatsby-plugin-google-gtag
```

2. Include the gatsby-plugin-google-gtag plugin in your gatsby-config. js file:

```
{
    resolve: `gatsby-plugin-google-gtag`,
    options: {
      trackingIds: [
        "GA-TRACKING_ID", // Your Measurement ID
      ],
      gtagConfig: {
```

```
                anonymize_ip: true
        },
    },
},
```

By including this plugin in your configuration, Gatsby will append the required Google Analytics script to the body of your application. The plugin comes with plenty of options, all of which can be found here: `https://www.npmjs.com/package/gatsby-plugin-google-gtag`.

It's important to draw attention to the `anonymize_ip` gtag configuration option. Anonymizing the IP is a legal requirement in some countries, such as Germany. Without any additional configuration, the plugin will automatically send a page view event whenever your site's route changes.

This is a great start, but you will most likely also want to track other events on your site. Let's look at how we can track custom events and outbound links.

Custom events

There are plenty more ways to track user engagement than just page views. Websites today are becoming more and more interactive, so being able to track interaction can be very useful. We can achieve this within Google Analytics by using custom events. Let's look at an example of using a button click.

Let's assume we have a simple button component:

```
import React from "react"
const Button = () => {
    return (
        <button>Click Me</button>
    )
}
export default Button
```

To track a click of this button, we can utilize the `gtag` function, which is exposed in the window via the plugin:

```
import React from "react"
const Button = () => {
    const track = (e) => {
        typeof window !== "undefined" &&
```

```
            window.gtag("event", "click", { /* Meta Data */ })
    }
    return (
        <button onClick={track}>Click Me</button>
    )
}
export default Button
```

In the previous code block, you can see that within `onClick`, we call a `track` function. This function calls the `window.gtag` function conditionally, if `window` is defined. We need to perform this check as the function does not work when it's rendered on the server side.

> **Important Note**
>
> This plugin is for production use only. This means that any events that take place while you are working on the project in development will not be tracked. To test that the plugin is working correctly, you will need to build and serve the site.

Now that we understand how custom events work, let's look at how we can track people leaving our site via outbound links.

Outbound links

It can be useful to understand where and when users navigate away from your site. Perhaps you reference another developer's site within a blog post and users leave to visit that site? To track this kind of outbound traffic, the `gatsby-plugin-google-gtag` plugin contains a ready-made component – `OutboundLink`. Let's see how we can use it:

```
import React from "react"
import { OutboundLink } from "gatsby-plugin-google-gtag"
const MyLink = () => {
    return (
        <OutboundLink href="https://sld.codes">Visit
        sld.codes.</OutboundLink>
    )
}
export default MyLink
```

As you should be able to see in this example, we can use the `OutboundLink` component as a direct drop-in replacement for an a tag. As its name suggests, you should only use this component for outbound links. If the link is internal, you should be using Gatsby's `Link` component.

Google Analytics is a great way to track your site's page analytics, but there is also a multitude of other tools you can use for this purpose. Let's look at an alternative – Fathom Analytics.

Using Fathom Analytics

Fathom Analytics is marketed as a privacy-focused alternative to Google. Google Analytics collects an abundance of data when users browse your site, but Fathom suggests that the information that is acquired is too much. Fathom collects only what they need to create their one-page stats dashboard. Unlike Google, Fathom is not free and starts at $14 per month. To start tracking data within our Gatsby site, we will need to obtain a site ID from Fathom Analytics. Let's do this now by following these steps:

1. Navigate to `https://usefathom.com` from your browser and create an account. You will need to sign up with a credit card, but you will receive a 7-day free trial.

2. Upon account creation, you should be prompted to create a new site:

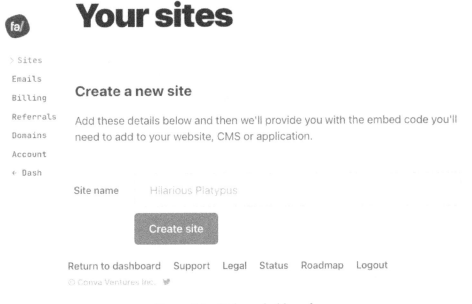

Figure 8.5 – Fathom dashboard

Give your site an appropriate name. A name such as **personal-website** would be appropriate. Click **Create site**.

3. Upon submission, you will be presented with your **Site ID** and embeddable code for a multitude of frameworks, including Gatsby:

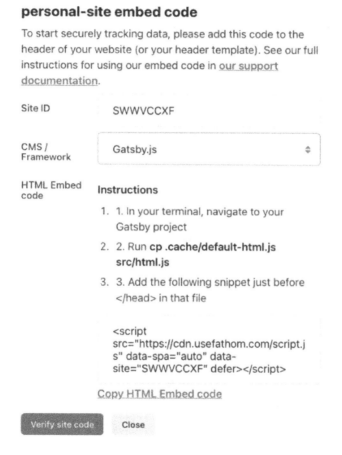

personal-site embed code

To start securely tracking data, please add this code to the header of your website (or your header template). See our full instructions for using our embed code in our support documentation.

Site ID   SWWVCCXF

CMS /   Gatsby.js   ⇕
Framework

HTML Embed   **Instructions**
code
1. 1. In your terminal, navigate to your Gatsby project
2. 2. Run **cp .cache/default-html.js src/html.js**
3. 3. Add the following snippet just before </head> in that file

<script src="https://cdn.usefathom.com/script.js" data-spa="auto" data-site="SWWVCCXF" defer></script>

Copy HTML Embed code

Verify site code   Close

Figure 8.6 – Fathom embed code

I suggest that you *do not* follow the instructions they provide you with for Gatsby. Why? Because their instructions suggest modifying the component that Gatsby uses to server render head and other parts of the HTML outside of Gatsby. There is no guarantee that Gatsby will keep this file consistent between versions, so meddling with it might make upgrading it difficult later down the line. Instead, simply make a note of your **Site ID**. Then, minimize your browser.

Now that we have retrieved our **Site ID**, let's turn our attention to our Gatsby site repository and start populating that Fathom dashboard with statistics:

1. Install the necessary dependencies:

    ```
    npm install gatsby-plugin-fathom
    ```

2. Include the `gatsby-plugin-fathom` plugin in your `gatsby-config.js` file:

    ```
    {
            resolve: 'gatsby-plugin-fathom',
            options: {
              siteId: 'FATHOM_SITE_ID'
            }
    }
    ```

3. Replace `FATHOM_SITE_ID` with the **Site ID** property that you retrieved via Fathom's website.

4. Build and serve your Gatsby site. `gatsby-plugin-fathom` is production-only, so we need to create and serve a production build to validate that it is working.

5. Once your site has loaded, navigate to it. Once it has been rendered, return to Fathom and click **Verify site code**. It should inform you that fathom is all hooked up!

Now, let's investigate how we can track more than page views.

Custom events (goals)

Like Google Analytics, Fathom also allow you to track custom events. Fathom refers to these events as "goals." To learn how we can track a goal, let's look at an example of using a button click.

First, we need to create an event via Fathom:

1. Navigate to your Fathom analytics dashboard. Under the **Events** section, click **Add event**.

2. Give your event a name and click **Create event**. Note your event code.

Now that we have an event code, let's use it! Let's assume we have a simple button component:

```
import React from "react"
const Button = () => {
    return (
        <button>Click Me</button>
    )
}
export default Button
```

To track a click of this button, we can utilize the useGoal function, which is exposed via the plugin:

```
import React from "react"
import { useGoal } from "gatsby-plugin-fathom"

const Button = () => {
    const handleGoal = useGoal("YOUR_EVENT_CODE")
    return (
        <button onClick={() => handleGoal(100)}>Click
            Me</button>
    )
}
export default Button
```

The useGoal hook exposes a function with a single argument, which is the value of your goal. Perhaps this is the purchase button, and you would like to log your revenue on your dashboard. If your goal has no value, set this parameter to 0.

Now that we understand how to track page analytics, let's look at how we can monitor our application for errors in production through application monitoring.

Monitoring the performance of your site

One of the hardest things to debug is user errors in production that you cannot seem to replicate on your machine. Without logs, this can be an impossible task. Luckily, some tools are dedicated to monitoring errors within your application and alert you when things do go wrong. One of the most popular tools out there for this purpose is Sentry.io.

Using Sentry.io analytics

Sentry.io is a full-stack error tracking system that supports a variety of desktop, browser, and server applications – including GatsbyJS! Sentry works by integrating with our site's logging infrastructure directly. Let's learn how we can implement Sentry so that we can monitor it for production errors:

1. Navigate to `https://sentry.io/signup/` from your browser and create an account.

2. Once you've logged in, create a new project by navigating to **Projects** and clicking **Create Project**.

3. Fill in the new project user interface, like so:

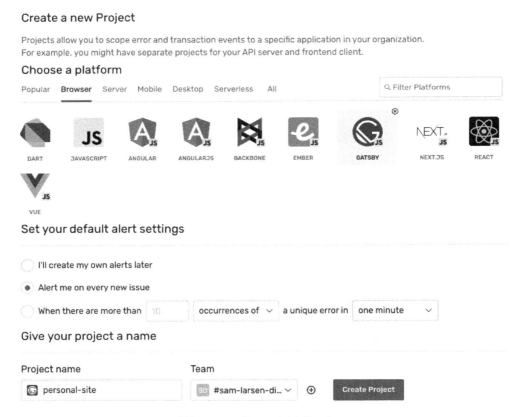

Figure 8.7 – Sentry initialization

Choose Gatsby as your platform. As your site is presumably going to be small at launch, I would suggest setting your default alert settings to **Alert me on every new issue**. Finally, give your project a name. Then, click **Create Project**.

Sentry, then, gives you a great step-by-step guide on how to set up Sentry within your Gatsby project. Let's reiterate these steps here:

1. Install the necessary dependencies:

    ```
    npm install --save @sentry/gatsby
    ```

2. Include the `@sentry/gatsby` plugin in your `gatsby-config.js` file:

    ```
    {
          resolve: "@sentry/gatsby",
          options: {
            dsn: "YOUR_DSN_NUMBER",
            sampleRate: 0.7
          },
    },
    ```

    `sampleRate` is the rate of error events that are sent. Sentry suggests `0.7` as the default value, which means that 70% of error events will be sent.

    A complete list of plugin options can be found here: `https://docs.sentry.io/platforms/javascript/guides/gatsby/configuration/options/`.

These few lines of code are all that is needed to have Sentry start tracking errors and the performance of your site in production. You can rest easy knowing that if your site visitors encounter errors, you will know about them instantly.

Summary

In this chapter, we learned about website analytics and how useful they can be to make our application perform at its best. We learned about the different types of data we can collect and the regulations we should follow when collecting a user's data. We implemented page analytics in two different ways – one by providing you with an abundance of data and the other by taking a more privacy-focused stance. Finally, we also implemented application monitoring using Sentry.io. You should now feel confident collecting website analytics.

In the next chapter, we will finally bring together everything we have learned in the last eight chapters and deploy our website.

9
Deployment and Hosting

In this chapter, we will finally take the project we have been working on and deploy it for the world to see! We will delve into the different types of builds that Gatsby creates and gain an understanding of how to debug common build errors. Following this, we will continue to learn how we can deploy them using a variety of different platforms. Additionally, we will discover how we can lock down access to our site by serving it up as part of an Express server.

In this chapter, we will cover the following topics:

- Understanding build types
- Common build errors
- Your pre-deployment checklist
- Platforms for deploying hybrid builds
- Platforms for deploying static builds
- Serving a Gatsby site with reduced user access

Technical requirements

To navigate this chapter, you will need to have completed *Chapter 8, Web Analytics and Performance Monitoring.*

The code that is presented in this chapter can be found at `https://github.com/ PacktPublishing/Elevating-React-Web-Development-with-Gatsby-4/ tree/main/Chapter09`.

Understanding build types

Gatsby version 4 introduced the ability for your website to be built in two different ways:

- As a **static build**: This creates all your pages at build time using Node.js. The resulting files are all static HTML, JavaScript, and CSS, which can be served entirely statically.

- As a **hybrid build**: This is a mixture of a static build combined with pages that are server-side rendered or have been created via deferred static generation.

When running `gatsby build`, Gatsby will inspect your site's content, and if possible, create a static build. However, if your site contains pages that are server-side rendered or have been created via deferred static generation, it will create a build that requires server-side code that runs on a Node.js server or via serverless functions. Builds of both types can be tested locally using the `gatsby serve` command.

Before deploying your build, it's worth ensuring that everything is working as it should locally. Now, let's take a moment to look at common build errors and learn how you can avoid them.

Common build errors

While working on our project, we mostly run the project in development mode. This is a great idea to ensure that the site also works with a production build by running the `gatsby build` command.

Sometimes, you might find that errors occur during the build process. So, let's talk about the most common issues and how we can fix them:

- The most common error that you'll come across is **window/document is not defined**. Node.js does not contain the `window` and `document` variables found in the browser. Therefore, while your site is being built, it is unable to access them. You can get around this issue in a couple of ways. You can perform a check to confirm that the variable is defined (for example, `typeof window !== undefined && yourFunction()`), or if appropriate, you can move the code into a `useEffect` hook.

- Ensure that all your components, your pages, and your `gatsby-browser.js` and `gatsby-ssr.js` files do not mix **ES5** and **ES6** syntax, as this can lead to builds crashing out.

- Take special care to ensure that all JavaScript files found within your `pages` directory are React components with a default export. Gatsby treats all JavaScript files as pages within this folder. If you have components or other utility functions within this directory, you will get an error that says **A page component must export a React component for it to be valid**. If you see this error, just move the files in question outside of the folder.

Now that we can build our site without issues, let's examine a practical checklist that we should run through before deploying our site.

Your pre-deployment checklist

Regardless of how you intend to deploy your site, there are a few steps you should follow on your local machine to ensure that your first deployment will run smoothly:

1. *Ensure any deployment platform plugins that are required have been installed.* A couple of the platforms we will look at have Gatsby plugins specifically for use with their product. By adding them to your Gatsby site, the platform is better able to understand your project and, as a result, build your site faster.

2. *Make sure your Gatsby site builds without an error.* Once the build has passed successfully, try running `gatsby serve` to ensure that you can use the site without issue.

3. *Ensure all your tests are passing.* Make sure that you have run your unit tests that we set up in *Chapter 7, Testing and Auditing Your Site*, using `npm run test`, and ensure that they are all passing.

4. *Take note of your Node.js version.* As of Gatsby version 4, your Node.js version should be 14 or higher. You'll want to ensure that the Node.js version matches your deployment platform so that you don't have compatibility issues. You can check this by running `node -v` in your terminal.

Now that we have completed our checklist, let's look at the various platforms we can deploy our site with, starting with those that support hybrid sites.

> **Important Note**
>
> It is recommended that you only deploy your site on one deployment platform and not multiple platforms. Managing multiple platforms when you can do the job with one is far easier for you to maintain. Try experimenting with all the options to find the best fit for your project.

Platforms for deploying hybrid builds

As hybrid sites require a Node.js server, we need to use platforms that can provision them. Hybrid sites are also very new to the Gatsby ecosystem. At the time of writing, the only stable option for hosting a hybrid build is Gatsby Cloud Hosting, so let's look at the platform next.

Deploying to Gatsby Cloud Hosting

Gatsby Cloud is a cloud platform that has been specifically designed and built for the Gatsby framework by the Gatsby organization. Because they focus on this framework, they excel at building technology that makes your builds run as fast as possible. This includes the following:

- **Incremental builds**: Gatsby Cloud observes the GraphQL data layer and identifies page dependencies. When you push changes to your code, it identifies the data layer changes and only rebuilds the pages that are dependent on that data. This can drastically speed up repeat builds – Gatsby says that incremental builds can be as much as 1,000 times faster than traditional builds.

- **Intelligent caching**: Special caching headers are sent to the browser when requesting your site. These are used to ensure that the browser does not re-download any content that has not changed between builds.

It should be noted that incremental builds are not available on the free tier of the platform. If you want to benefit from them, you'll need to upgrade.

Now that we understand the benefits of using the platform, let's look at how we can deploy our site to the platform.

> **Quick Note**
> The process for deploying a hybrid site and a static site is the same on the Gatsby platform, so these instructions will work in both cases.

Perform the following steps to deploy your site to the Gatsby Cloud platform:

1. Install the Gatsby Cloud plugin:

    ```
    npm install --save gatsby-plugin-gatsby-cloud
    ```

    Here, we are installing the Gatsby Cloud plugin. This adds basic security headers during the build for the Gatsby Cloud platform.

2. Include the `gatsby-plugin-gatsby-cloud` plugin in your `gatsby-config.js` file:

    ```
    module.exports = {
      // rest of config
      plugins: [
        `gatsby-plugin-gatsby-cloud`,
        // other plugins
      ]
    }
    ```

3. Commit and push all changes to your chosen Git repository.

4. Open a browser and navigate to `https://www.gatsbyjs.com/products/cloud`. Click on **Get Started**.

5. Sign up to the platform by populating the form with your name, email, and country of residence:

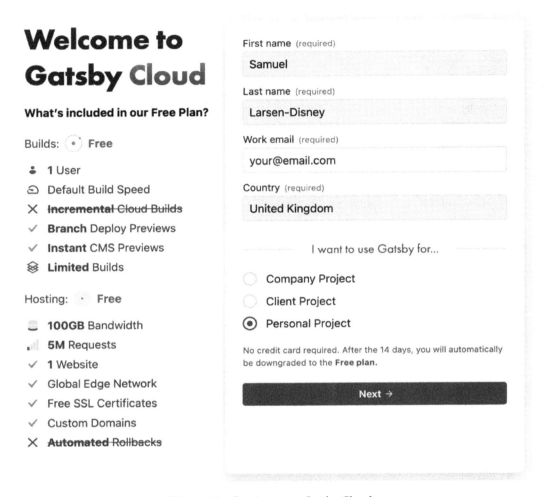

Figure 9.1 – Signing up to Gatsby Cloud

6. Select your **VCS** (**version control system**) provider, log in, and approve the requested permissions that the Gatsby Cloud platform requires in order to integrate with it:

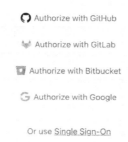

Figure 9.2 – The Gatsby Cloud VCS Provider authorization step

7. Upon being redirected to Gatsby Cloud, you will be asked whether you would like to trial a 14-day upgrade. This is up to you.

8. Then, you will be navigated to your dashboard, which will be empty, as we have not set up any sites yet. Let's add our site now by clicking on **Add Site**.

9. Select **Import from a Git Repository**, and click on **Next**.

10. Select your Git provider from the list, followed by the organization and repository name. If, for some reason, this list has not been populated, ensure that you have given Gatsby the relevant permissions to read from your Git repositories.

11. Following this, you will need to provide your site details, including a base branch and base directory. Your base directory should point to the root of the Gatsby project within the repository – this is most likely the root directory or /. Click on **Next**.

12. You will then be presented with optional integrations for your site. These integrations can help your CMS communicate with Gatsby Cloud. When you make a change to your CMS, you can see a preview of how that content will look via Gatsby Cloud. If you desire to do this, you can click on **Connect** next to the CMS platform you are using and follow the steps; otherwise, you can click on **Skip this Step**.

13. Finally, you will be asked to add any environment variables that your site needs to build. Gatsby scans your site's integrations and plugins to help fill in the environment variables it thinks you need. Be sure to cross-check this with your local .env file to ensure you have everything that is required.

14. Click on **Create site**. This will prompt Gatsby to start building your site for the very first time:

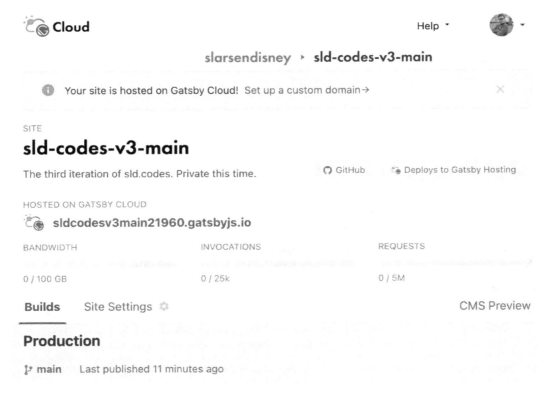

Figure 9.3 – The Gatsby Cloud site dashboard

Once the build has been completed, you can see the deployed site live by following the purple hyperlink underneath the **HOSTED ON GATSBY CLOUD** heading in the preceding screenshot.

With every subsequent push to the base branch, Gatsby Cloud will build and deploy the change automatically.

Now that we understand how to deploy a hybrid build, let's look at the options we have for deploying static builds.

Platforms for deploying static builds

As static builds are a far more common and predictable format, there are plenty more options for where you can host them. We have already looked at Gatsby Cloud, which can deploy static sites in the same way as it does hybrid. Now, let's look at three other platforms – Netlify, Render, and Firebase.

Deploying to Netlify

Netlify is the deployment platform used for over 500,000 websites. It is popular among developers for its ease of use. It also provides a free **Secure Sockets Layer** (**SSL**). Let's learn how we can deploy our site with Netlify:

1. Install the Netlify plugin:

    ```
    npm install --save gatsby-plugin-netlify
    ```

    Here, we are installing the Netlify plugin, which adds basic security headers during the build for the Netlify platform.

2. Include the `gatsby-plugin-netlify` plugin in your `gatsby-config.js` file:

    ```
    module.exports = {
      // rest of config
      plugins: [
        `gatsby-plugin-netlify`,
        // other plugins
      ]
    }
    ```

3. Commit and push all changes to a Git repository.

4. Navigate to `https://app.netlify.com/signup` in your browser.

5. Sign up by logging in with the third-party login details provided by your VCS, and approve the requested permissions that the Netlify platform requires to integrate with it.

6.   Click on **Create New Site** from your dashboard:

Create a new site

From zero to hero, three easy steps to get your site on Netlify.

1. Connect to Git provider   2. Pick a repository   3. Site settings, and deploy!

Continuous Deployment

Choose the Git provider where your site's source code is hosted. When you push to Git, we run your build tool of choice on our servers and deploy the result.

You can unlock options for self-hosted GitHub/GitLab by upgrading to the Business plan.

| GitHub | GitLab | Bitbucket |

Figure 9.4 – The Netlify new site page

Select the Git provider where your repository is stored. Then, pick the repository that you would like to build.

7.   You can leave the **Owner** option in its default setting. However, make sure that the deploy branch matches your site's main production branch:

Create a new site

From zero to hero, three easy steps to get your site on Netlify.

1. Connect to Git provider   2. Pick a repository   3. Site settings, and deploy!

Site settings for slarsendisney/sld-codes-v3

Get more control over how Netlify builds and deploys your site with these settings.

Owner

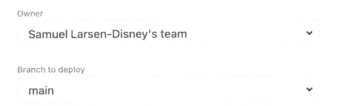

Samuel Larsen-Disney's team

Branch to deploy

main

Figure 9.5 – The Netlify site creation settings

8. Finally, provide the **Build command** and **Publish directory** details for the site, which should be `npm run build` and `public`, respectively:

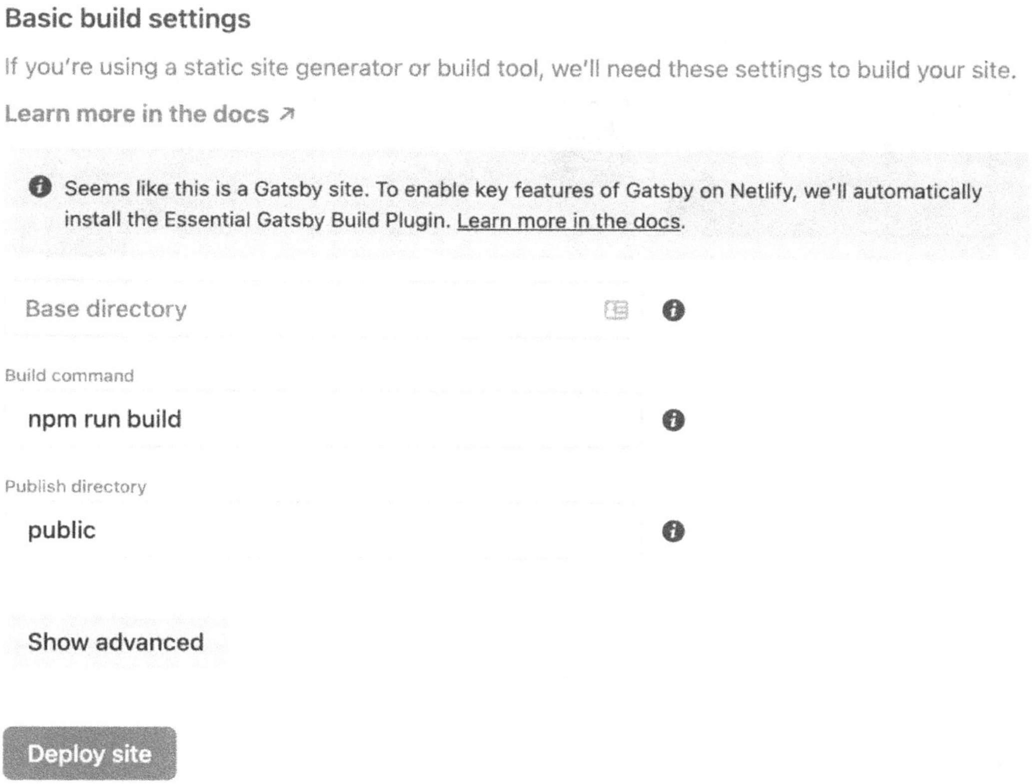

Basic build settings

If you're using a static site generator or build tool, we'll need these settings to build your site.

Learn more in the docs ↗

> ⓘ Seems like this is a Gatsby site. To enable key features of Gatsby on Netlify, we'll automatically install the Essential Gatsby Build Plugin. <u>Learn more in the docs</u>.

Base directory   ▣   ⓘ

Build command

npm run build   ⓘ

Publish directory

public   ⓘ

Show advanced

Deploy site

Figure 9.6 – The Netlify site creation build settings

9.   Clicking on **Deploy site** will start the build process.

10. While your site is building, take note of the URL, in blue, at the top of the dashboard. This is where your site will be deployed:

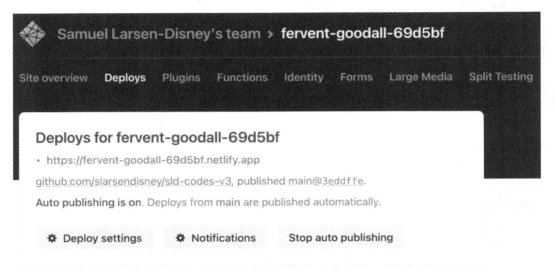

Figure 9.7 – The Netlify site dashboard

If everything goes well, your site should be deployed after a few minutes. With every subsequent push to the base branch, Netlify will build and deploy the change automatically.

Now that we understand how to deploy with Netlify, let's look at another alternative – Render.

Deploying to Render

Render is a cloud platform that can build and run Gatsby websites with free SSL and a global CDN. Let's learn how we can deploy our site with Render:

1. Commit and push all changes to a Git repository.

2. Navigate to `https://dashboard.render.com/register`, and then create an account.

3. From the dashboard, click on **New**:

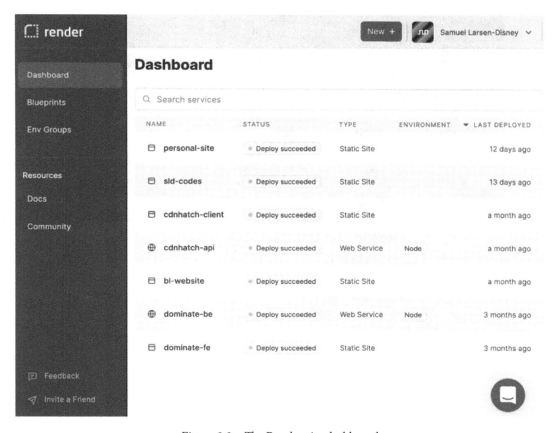

Figure 9.8 – The Render site dashboard

4. Select **Static Site**.

5. At this point, you will be asked to present a repository. However, as you have not connected Render to your VCS, the list will be empty. Click on the hyperlink for your VCS and proceed to connect Render to that system by following the UI journey from that third party. In the case of GitHub, it will look something like this:

Figure 9.9 – GitHub's third-party installation

6. The list should now be populated with your repositories. Select the one containing your Gatsby site.

7. Next, configure your site settings:

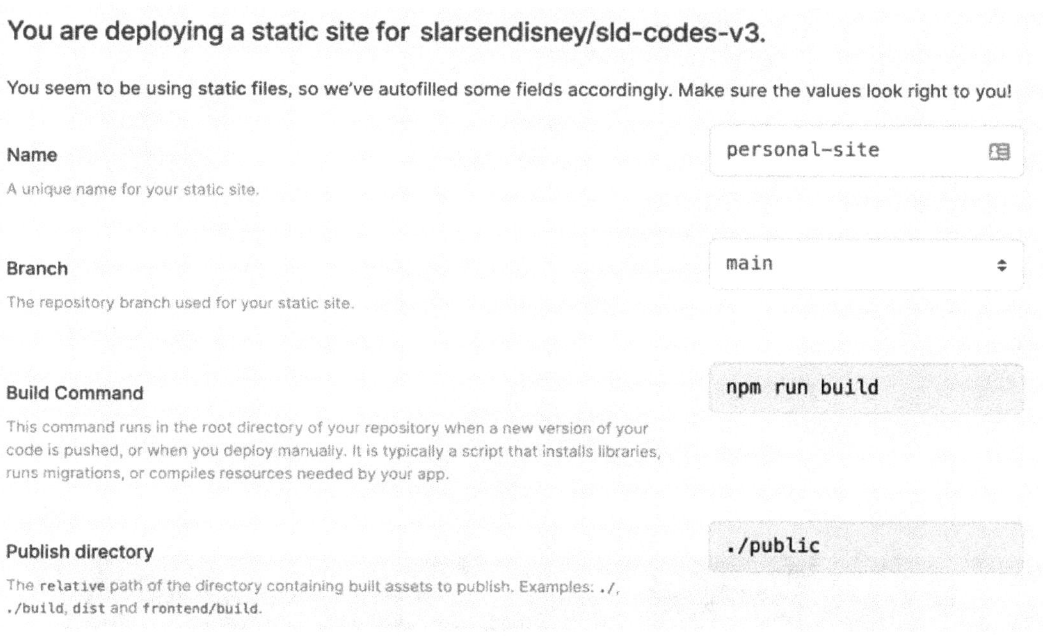

You are deploying a static site for slarsendisney/sld-codes-v3.

You seem to be using static files, so we've autofilled some fields accordingly. Make sure the values look right to you!

Name

`personal-site`

A unique name for your static site.

Branch

`main`

The repository branch used for your static site.

Build Command

`npm run build`

This command runs in the root directory of your repository when a new version of your code is pushed, or when you deploy manually. It is typically a script that installs libraries, runs migrations, or compiles resources needed by your app.

Publish directory

`./public`

The relative path of the directory containing built assets to publish. Examples: ./, ./build, dist and frontend/build.

Figure 9.10 – The Render site settings

The **Name** field can be anything you like, and the **Branch** field should be the branch of your repository that you would like to be deployed. Additionally, **Build Command** should be npm run build, and **Publish directory** should be ./public.

8. Click on **Create Static Site**.

9. While your site is building, take note of the URL, in blue, at the top of the dashboard. This is where your site will be deployed:

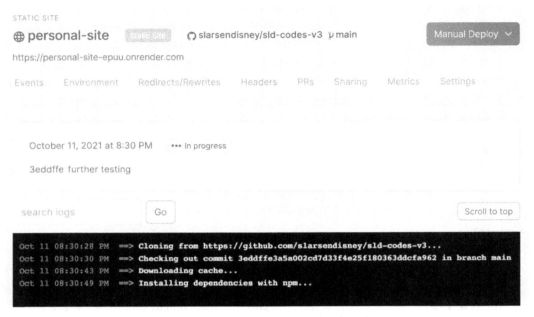

Figure 9.11 – The Render site dashboard

If everything goes well, your site should be deployed after a few minutes. Check the URL to be sure. With every subsequent push to the base branch, Render will build and deploy the change automatically.

Now that we understand how to deploy with Render, let's look at another alternative – Firebase.

Deploying to Firebase

Firebase is Google's mobile development application. It allows you to focus on the frontend of your application by allowing you to manage your backend infrastructure through a no-code/low-code development UI. Firebase has a large number of features, including real-time databases, machine learning, Cloud Functions authentication, and – the feature we will be focusing on – hosting. Let's learn how we can deploy our site with Firebase:

1. Navigate to `https://console.firebase.google.com`, and sign in with a Google account.

2. Once logged in, you will be directed to the Firebase console. From there, click on **Add Project**.

3. You will be prompted to give your project a name. Once entered, take note of your project ID, and click on **Continue**:

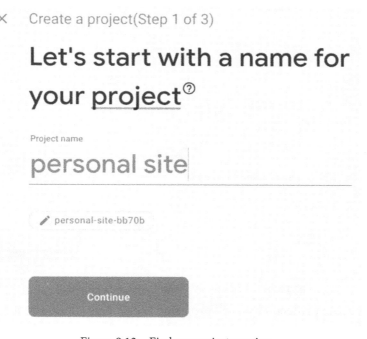

Figure 9.12 – Firebase project naming

4. At this point, you can optionally set up Google Analytics for your project. If you added Google Analytics to your site as part of *Chapter 8*, *Web Analytics and Performance Monitoring*, then do not set this up again here:

Figure 9.13 – Setting up Firebase project analytics

5. Click on **Create project** – this will provision the Google Cloud services that are required for your project. Now we have set up everything we need within the Firebase platform and can return to the code.

6. Install the Firebase CLI:

```
npm install -g firebase-tools
```

This package allows us to integrate local projects with the Firebase platform. We can use the -g command to install it globally.

7. Run the `firebase login` command:

```
firebase login
```

This will open a browser window prompting you to log in with a Google account. Log in with the Google account that you signed up to Firebase with.

8. Once complete, return to your Gatsby project's root directory and run the following:

```
firebase init
```

This will trigger the Firebase initialization UI within our Gatsby project and present you with the following:

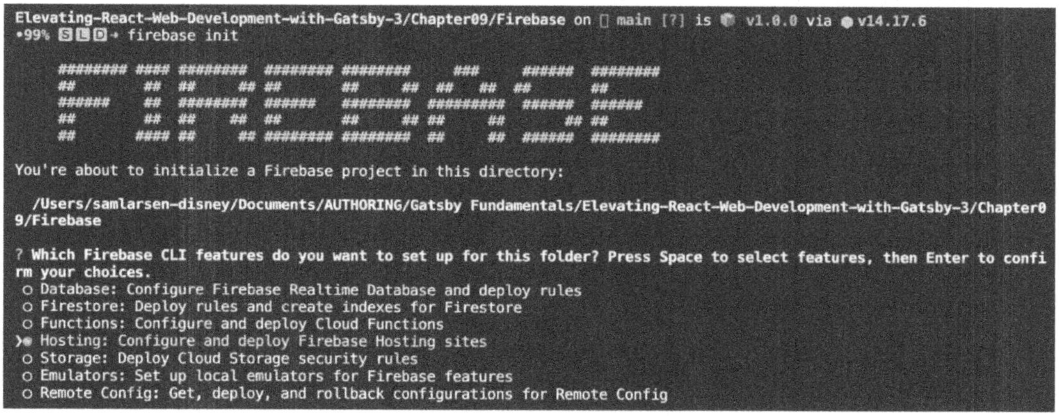

```
Elevating-React-Web-Development-with-Gatsby-3/Chapter09/Firebase on  main [?] is  v1.0.0 via  v14.17.6
•99%  firebase init

        ######## #### ####### ######## #######    ###    ##### #######
        ##    ## ##  ##    ## ##    ## ##    ##  ## ##  ## ## ##    ##
        ######   ##  ######## #####   ######## ######## ##### #####
        ##       ## ## ##  ## ##      ##    ## ##    ## ## ## ##  ##
        ##       #### ##   ## ## ######## ####### ##     ## ##### #######

You're about to initialize a Firebase project in this directory:

  /Users/samlarsen-disney/Documents/AUTHORING/Gatsby Fundamentals/Elevating-React-Web-Development-with-Gatsby-3/Chapter0
9/Firebase

? Which Firebase CLI features do you want to set up for this folder? Press Space to select features, then Enter to confi
rm your choices.
 o Database: Configure Firebase Realtime Database and deploy rules
 o Firestore: Deploy rules and create indexes for Firestore
 o Functions: Configure and deploy Cloud Functions
>● Hosting: Configure and deploy Firebase Hosting sites
 o Storage: Deploy Cloud Storage security rules
 o Emulators: Set up local emulators for Firebase features
 o Remote Config: Get, deploy, and rollback configurations for Remote Config
```

Figure 9.14 – Firebase CLI project initialization

Within this project, we are only using hosting, so press the down arrow key until **Hosting** is selected. Then, hit the spacebar to select it followed by *Enter* to confirm your choice.

9. Firebase will then ask you which Firebase project to associate with this directory:

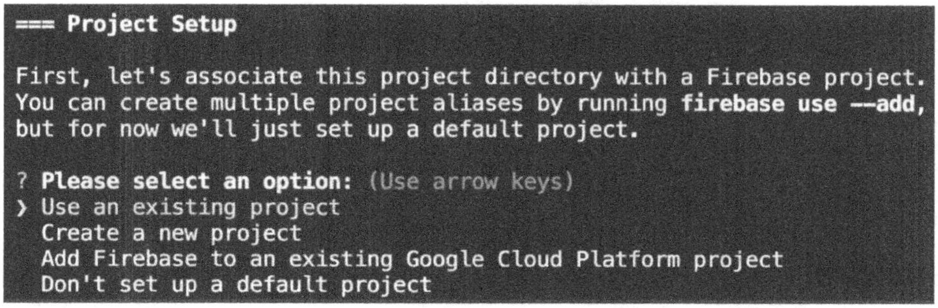

```
=== Project Setup

First, let's associate this project directory with a Firebase project.
You can create multiple project aliases by running firebase use --add,
but for now we'll just set up a default project.

? Please select an option: (Use arrow keys)
> Use an existing project
  Create a new project
  Add Firebase to an existing Google Cloud Platform project
  Don't set up a default project
```

Figure 9.15 – Setting up the Firebase CLI project

We have already created a Firebase project, so ensure **Use an existing project** is selected and hit *Enter*.

10. Use the up and down arrow keys to select the project ID that we created in *Step 3* (the project name should be visible in brackets next to the ID). Then, hit *Enter*.

11. Tell Firebase where to find the static build during the hosting setup:

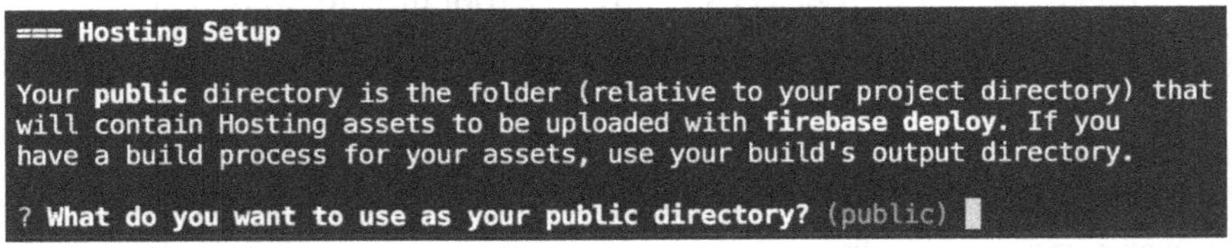

```
=== Hosting Setup

Your public directory is the folder (relative to your project directory) that
will contain Hosting assets to be uploaded with firebase deploy. If you
have a build process for your assets, use your build's output directory.

? What do you want to use as your public directory? (public)
```

Figure 9.16 – Setting up Firebase CLI hosting

By default, Firebase uses the `public` directory, so we can hit *Enter* without changing this.

12. Then, it will ask you whether you would like it to configure your application as a single-page app. Type in n and hit *Enter*.

13. Finally, it will ask whether you want to set up automatic deploys with GitHub. Type in n and hit *Enter*. You can change this in the future if needed, but for now, we will focus on manual deployments.

14. We now have everything in place ready to deploy to Firebase. Run the following command:

```
gatsby build && firebase deploy
```

As we already know, `gatsby build` will create a production-ready build of our site. Then, the `firebase deploy` command will take our build and upload it to the Firebase platform, ready to be served to site visitors:

```
=== Deploying to 'personal-site-bb70b'...

i  deploying hosting
i  hosting[personal-site-bb70b]: beginning deploy...
i  hosting[personal-site-bb70b]: found 116 files in public
✔  hosting[personal-site-bb70b]: file upload complete
i  hosting[personal-site-bb70b]: finalizing version...
✔  hosting[personal-site-bb70b]: version finalized
i  hosting[personal-site-bb70b]: releasing new version...
✔  hosting[personal-site-bb70b]: release complete

✔  Deploy complete!

Project Console: https://console.firebase.google.com/project/personal-site-bb70b/overview
Hosting URL: https://personal-site-bb70b.web.app
```

Figure 9.17 – Deploying the Firebase CLI

At the end of the Firebase deployment, it will log a **Hosting URL** to the terminal. Navigate to this link in a browser to see your deployed application.

> **Quick Tip**
>
> As you might have noticed from these instructions, Firebase is the only platform in this list that does not require you to push your code to a VCS. If you have a project where you do not wish to use a VCS, this is a great choice. It's important to note that, unlike the other platforms, Firebase will not automatically deploy your project unless it has been set up as part of a deployment pipeline.

We have now looked at a multitude of different ways to deploy our site onto the internet. If you have completed any of the implementations discussed in the previous sections, you should be able to send a friend the site URL, and they should be able to see it. However, what if you don't want your site to be visible to everyone but only a selected few? Next, let's look at how we can reduce the level of access to our site for when the situation requires it.

Serving a Gatsby site with reduced user access

You might be asking yourself why you would ever want to reduce the access to your site. One word – security. In all the examples we have seen so far, our site is public and out there on the internet for all to see, but what if you are building an application that is only for a selected group of people? Perhaps it's portfolio work that you want to have locked behind a password or an onboarding application that should only be available to colleagues at a specific company. We can achieve functionalities such as these using most backend web applications.

> **Important Note**
>
> This type of authentication is not to be confused with that of *Chapter 11, Creating Authenticated Experiences*. Here, we are restricting access to the entirety of the site unless you have been approved. In *Chapter 11, Creating Authenticated Experiences*, access is only partially restricted, as we allow users to visit parts of the application without logging in.

As an example, let's explore how we can use Express to introduce a password login to our site:

1. Install the dependencies:

    ```
    npm i express express-basic-auth
    ```

    We will be using Express as our backend and `express-basic-auth` to implement HTTP basic authorization as middleware.

2. At the root of your Gatsby project, create a `server.js` file with the following:

    ```js
    const express = require("express");
    const app = express();
    const basicAuth = require("express-basic-auth");
    const port = 3000;

    app.use(
      basicAuth({
        challenge: true,
        users: { admin: "testing" },
      })
    );

    app.use(express.static(`${__dirname}/public`));

    app.listen(port, () => {
      console.log(`Example app listening at
        http://localhost:${port}`);
    });
    ```

First, we create an `express` app. Then, we instruct it to use the `express-basic-auth` middleware. You will see that we are passing an object that instructs the middleware to challenge the user. When a user navigates to the site, before seeing any content, they will be prompted with the following dialog box:

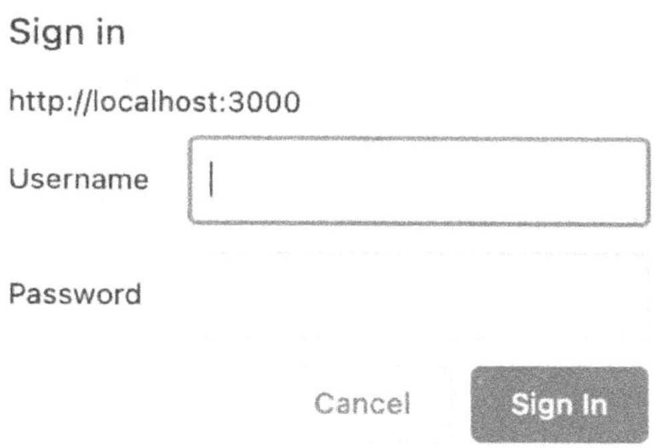

Sign in

http://localhost:3000

Username

Password

Cancel   **Sign In**

Figure 9.18 – Basic auth challenge dialog box

They will only be allowed onto the site if the credentials they provide match those in the users object provided to basicAuth.

Assuming they successfully pass this middleware check, we then allow them to view the static content of our site using the express.static() method.

3. Modify your scripts in package.json to include a start:server script:

```
"scripts": {
  ...
    "start:server": "node server.js"
  },
```

This script will run our server.js file using Node.js.

4. We now have everything in place to try our server:

```
gatsby build && npm run start:server
```

This will build your Gatsby project and then run your server code, which will serve up your Gatsby build content. If all has gone well, you should be able to visit localhost:3000 and see this implementation working. Upon entering the username and password that has been specified on the server, you should be able to see your Gatsby application.

All the other static deployment methods we have looked at within this chapter have assumed that the Gatsby project is being hosted on its own dedicated server, but sometimes, you don't always have the luxury of multiple servers. This example is also a great demonstration of how you can combine backend and frontend code on a single server. You could use a similar approach to lock down your site to certain IP ranges. For instance, we could expand upon this Express server to serve API endpoints alongside our Gatsby project within the same repository.

> **Quick Tip**
> You might be wondering how to deploy a site using this functionality – deploying Express servers is beyond the remit of this book, but platforms that support this include Heroku, Render, and Google Cloud.

Now, let's take a moment to summarize what we have learned in this chapter.

Summary

In this chapter, we investigated the builds that a Gatsby project can create and the differences between them. We looked at common errors that crop up during build time and how we can debug them. We learned how we can deploy hybrid builds using Gatsby Cloud and how we can deploy static builds with Netlify, Render, and Firebase. Additionally, we discovered how we can lock down access to our site by serving it up as part of an Express server. You should now feel comfortable with the process of taking your site live.

In the next chapter, we will start looking at more advanced concepts. We will begin by learning about Gatsby's plugin creation.

Part 3: Advanced Concepts

By now, you should have an understanding of how to make a standard static site live. In this part, we cover some more advanced techniques to handle use cases that are slightly less common for Gatsby sites.

In this part, we include the following chapters:

- *Chapter 10, Creating Gatsby Plugins*
- *Chapter 11, Creating Authenticated Experiences*
- *Chapter 12, Using Real-Time Data*
- *Chapter 13, Internationalization and Localization*

10
Creating Gatsby Plugins

In this chapter, we will look at Gatsby's plugin ecosystem. We'll start by learning how to make our Gatsby site more modular as it grows. We will then create our first source plugin to fetch data from GitHub. We will also create our first theme plugin to create events pages for our website. Finally, we will learn how to share our plugins with the world via Gatsby's plugin ecosystem.

In this chapter, we will cover the following topics:

- Understanding Gatsby plugins
- Introducing local plugin development
- Creating source plugins
- Creating theme plugins
- Contributing to the plugin ecosystem

Technical requirements

To complete this chapter, you will need to have completed *Chapter 9, Deployment and Hosting*. You will also need a GitHub account.

The code for this chapter can be found at `https://github.com/PacktPublishing/Elevating-React-Web-Development-with-Gatsby-4/tree/main/Chapter10`.

Understanding Gatsby plugins

By this stage in this book, you should have all the tools you need to get a Gatsby site into production. In this chapter, we are going to go one step further and talk about creating reusability across multiple Gatsby sites using something called **Gatsby plugins**. Gatsby plugins are node packages that abstract common site functionality that utilizes Gatsby APIs. By bundling functionality into a plugin, you can source data, create pages, implement SEO, and so much more with just a few lines. Gatsby plugins also act as a way to modularize larger sites into more manageable chunks of functionality.

The two most common types of plugins are as follows:

- **Gatsby Source Plugins**: Source plugins allow you to gather data from a data source and ingest it into Gatsby's GraphQL data layer. You could source data from anywhere, such as APIs, RSS feeds, or CMSes, as we did in *Chapter 3, Sourcing and Querying Data (from Anywhere!)*. Once data has been ingested into your GraphQL data layer, you can query it from within your `gatsby-node.js` file, as well as within your pages.

- **Gatsby Theme Plugins**: Theme plugins focus more on the user interface of your application. Often, theme plugins contain code that creates pages of a site, such as an FAQ section. They act to split your Gatsby site into smaller manageable projects, which can be very useful when you have multiple teams working on the same site.

These two types can be identified by the plugin name, which will either begin with `gatsby-source` or `gatsby-theme`. While these two types are the most common, they are not the only types. Plugins that encapsulate any other functionality have a plugin name that begins with `gatsby-plugin`.

Before we dive in and start creating plugins, let's learn about local plugin development so that we can avoid common pitfalls.

Introducing local plugin development

Local plugin development begins with a new folder called `plugins`, which you need to create within your root directory. This is the folder that will house the plugins we create. When you add a plugin to your Gatsby config, Gatsby first looks within your `node_modules` folder. If it cannot find a plugin there, it will check within this local `plugins` folder. If it finds a plugin here with the same name within its `package.json` file, it will use it.

As you may have guessed by the mention of a `package.json` file, plugins come in the form of npm packages. npm packages take care of their dependencies, so it is important that, when you're installing packages for use in a plugin, you make sure that you open the terminal within the plugin's folder and not the root directory. Otherwise, your site and plugin dependencies may be inaccurate.

> **Quick Tip**
>
> If you don't have any intention of ever sharing the plugins you create, you can choose to install dependencies that your plugins require in the root directory instead. This can be easier to manage if you prefer having one source of truth for your dependencies. But be careful – *if you think the plugin could be shared at any point, do not do this*, as you will have to manually sort through your dependencies and find those that the plugin requires.

While creating local plugins, you may find that your code does not behave like the rest of your project. Let's look at how we can debug common issues.

Debugging local plugins

Gatsby does not treat the local plugins folder the same as the rest of the code base. Changes to pages, templates, and configs may not necessarily appear while hot reloading. Here are a couple of tips to make your life a little easier:

- If you make a change and do not see it reflected, even after restarting the server, try clearing the cache using `gatsby clean`. Gatsby caches plugin data in the `.cache` folder. To make itself faster, Gatsby uses this cache.

- If you are unsure whether your plugin is even being run, try adding the following command to your plugin's `gatsby-node.js` file:

```
exports.onPreInit = () => console.log("Plugin
  Started!")
```

This command will run first during Gatsby's execution. If Gatsby is aware of your plugin, you will see **Plugin Started!** logged to the console.

Now that we know when it's a good idea to make plugins, let's learn how we can create them.

Creating source plugins

As we mentioned in the *Understanding Gatsby plugins* section, source plugins are those that allow us to ingest data from a new source into our GraphQL layer. By creating a source plugin, we abstract the logic to source this data away from our site so that we can reuse it across multiple Gatsby projects if we want to. To understand how source plugins work, let's build one together. Let's source our total contributions from GitHub so that we can display them on our **about** page:

1. The first thing we need to be able to do to pull data from GitHub is use an access token. Navigate to `https://github.com/settings/tokens/new`.

2. Write a **Note** to help you identify your access token later:

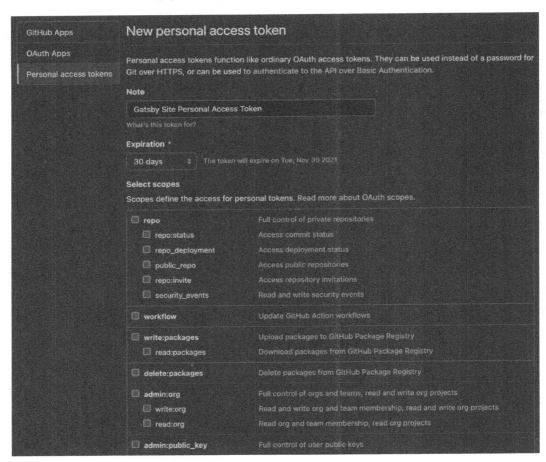

Figure 10.1 – GitHub personal access token generation

3. Change the **Expiration** property to your desired length. Once the length of time has been selected, the token will be deleted and no longer work. If you prefer that it doesn't expire, you can select **No Expiry** from this list.

4. Scroll down the list and check **read:user**:

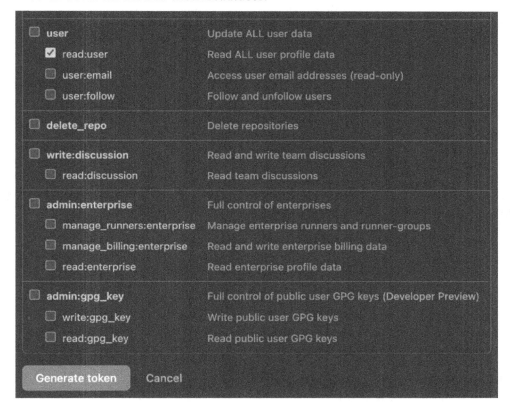

Figure 10.2 – GitHub personal access token generation (continued)

5. Click **Generate token**.

6. On the next screen, you will be presented with your access token – make a note of this immediately as you will not be able to see it again.

> **Important Note**
>
> If you ever lose your access token, you will not be able to see it again. GitHub does this to prevent your key being used by someone else for malicious purposes. In an instance where you do lose your key, you will have to create a new one with the same scopes and replace the token wherever it was being used.

7. Create a `.env` file in your root directory and add the following line:

```
GITHUB_PROFILE_BEARER_TOKEN=your-token-here
```

You may already have a `.env` file within your project at this point. If this is the case, simply append the preceding code block line to that file.

8. Ensure that `dotenv` is installed as a dependency at the root of your project. If it is not, run the following command:

```
npm i dotenv
```

9. Create a new folder called `gatsby-source-github-profile` in your `plugins` folder.

10. Open a terminal in the `gatsby-source-github-profile` folder and run the following command:

```
npm init -y
```

This initializes an npm package for our plugin.

11. Install the `node-fetch` package:

```
npm i node-fetch@2.6.5
```

The `node-fetch` package brings the `fetch` browser API to node. I've used it in this example as I suspect most of you will be familiar with `fetch`, as this book is aimed at React developers.

> **Important Note**
>
> Node Fetch is ESM only from version 3.0. This means it will not play nicely with the ES5 format that's being used in our Gatsby configuration files. The maintainers suggest using version 2.6.5 in our case.

12. Create a `gatsby-node.js` file in your `gatsby-source-github-profile` folder and add the following code to it:

```
const fetch = require("node-fetch");
const crypto = require("crypto");
/*
  Code added here in the next step
*/
```

Here, we are importing our most recent install, `node-fetch`, and the `crypto` library (which comes with node) into our project. `crypto` provides cryptographic functionality, which we will be using later in this file.

13. Under your imports, add the following code:

```
exports.sourceNodes = async ({ actions },
  configOptions) => {
  const { createNode } = actions;
  /*
    Code added here in the next step
  */
};
```

Here, we are utilizing the `sourceNodes` Gatsby node API. As its name suggests, we will add code here that sources our data and then creates nodes using the `createNode` action. You may have also noticed that we are passing `configOptions` into this as an argument. This object gives us access to any of the options we provide to the plugin when we use it in our `gatsby-config.js` file. We are going to be passing our access token and username as options.

> **Quick Tip**
>
> To improve the understandability of this file, it's been broken down into its parts. If you are finding it hard to follow, you can see the file in its entirety within the repository listed in the *Technical requirements* section of this chapter.

14. Create a `POST` request, like the following, inside `sourceNodes` for the GitHub API:

```
const headers = {
  Authorization: 'bearer ${configOptions.token}',
};
const body = {
  query: 'query {
          user(login: "${configOptions.username}") {
            contributionsCollection {
              contributionCalendar {
                totalContributions
              }
            }
```

```
            }
          }',
  };
  const response = await
  fetch("https://api.github.com/graphql", {
    method: "POST",
    body: JSON.stringify(body),
    headers: headers,
  });

  const data = await response.json();
  /*
    Code added here in the next step
  */
```

We use node-fetch to make the POST request to the GitHub API. We provide it with token authentication in the request header. Here, you can see we are using the token that's provided within configOptions. Like Gatsby, the GitHub API uses GraphQL. As with any GraphQL API, to select which data we want from GitHub, we have to pass a query into the body of our request. The query that's defined in body retrieves the total contributions for a given username (in this case, yours!). We pass our username in from configOptions.

15. Add the following code after your request:

```
const { contributionsCollection } = data.data.user;
  const totalContributions =
    contributionsCollection.contributionCalendar.
totalCont
  ributions;
  createNode({
    totalContributions: Number(totalContributions),
    id: "Github-Contributions",
    internal: {
      type: 'GitHubContributions',
      contentDigest: crypto
        .createHash('md5')
        .update(
          JSON.stringify({
```

```
        totalContributions,
      })
    )
    .digest('hex'),
  description: 'Github Contributions Information',
  },
});
```

Here, we deconstruct the data from our request to receive the total contributions. Then, we utilize the `createNode` function to add this data to our GraphQL data layer. Let's break down the object I am passing to the function:

a. `totalContributions`: The first key value in the object is the value of the total contributions. This is the variable we will query for later when we try to retrieve this information on our pages.

b. `id`: Each node must have a globally unique ID. Because there is a single instance of this node type, we can just use the`"Github-Contributions"` string.

c. `internal.type`: A globally unique type that we can use to identify this data source.

d. `internal.contentDigest`: This field helps Gatsby avoid regenerating nodes when they haven't changed. While creating the node if this field remains constant, it won't regenerate. So, we need to make sure that if our total contributions change, so too does this `contentDigest`. To do that, I am using the `crypto` library to create an `md5` hash of our total contributions. This might seem a little overkill in this particular instance, but it works well if the amount of data on a node is more than one key-value pair, as you can just add them to the object that's being passed to `JSON.stringify`.

e. `internal.description`: This field allows us to describe the source type, which is helpful if we are confused about what this source is at any point. This field is not required but is nice to have. Our plugin is now ready to be used – the process from this point is the same as it is for a plugin that's been installed via npm.

16. Navigate to your `gatsby-config.js` file at the root of your project and add the following code:

```
require("dotenv").config({
  path: '.env',
});
```

```
module.exports = {
  // rest of config
  plugins: [
    {
      resolve: 'gatsby-source-github-profile',
      options: {
        token:
          process.env.GITHUB_PROFILE_BEARER_TOKEN,
        username: "your-github-username-here",
      },
    },
    // other plugins
  ]
}
```

Note that we are passing in the options to the plugin that we utilized in the plugin's `gatsby-node.js` file. We source the token from our `.env` file. You can pass your GitHub username in as plain text as this is public information.

Quick Note

You may be tempted to try adding someone else's username here instead of your own, but this will cause the fetch request to fail as your access token does not have permission to retrieve another user's data.

---

17. Start your development server. Navigate to `http://localhost:8000/_graphql` – you should be able to query your total contributions with the following query:

```
query Contributions {
  gitHubContributions {
    totalContributions
  }
}
```

18. Let's add this new source of data to our `about` page:

```
export default function About({ data }) {
  const {
    markdownRemark: { html },
```

```
      gitHubContributions: { totalContributions }
    } = data;
    return (
      <Layout>
        <div className="max-w-5xl mx-auto py-16 lg:py-24
        text-center">
        <div dangerouslySetInnerHTML={{ __html: html
          }}></div>
          <p>In the last year I have made
            <b>{totalContributions}</b> on Github.</p>
        </div>
      </Layout>
    );
}


export const query = graphql'
  {
    markdownRemark(frontmatter: { type: { eq: "bio" } })
  {
      html
    }
    gitHubContributions {
      totalContributions
    }
  }
';
```

We have updated the page query so that it includes the new source. We can then access it within the page via the data prop and render it to the screen, as shown in the highlighted section of code.

Congratulations – you've just built your first local plugin. You could replicate the methods outlined here to fetch data from another API. So, at this point, we can create source plugins with ease, but what about theme plugins?

Creating theme plugins

As we have discovered, theme plugins are all about adding visual elements to our Gatsby site. Theme plugins are unique in that they have to contain a `gatsby-config.js` file. To better understand theme plugins, let's look at the most minimal of examples. Let's use a plugin to add a simple sample page to our site:

1. Create a new folder called `gatsby-theme-sample-page` in your `plugins` folder.

2. Open a terminal in the `gatsby-theme-sample-page` folder and run the following command:

    ```
    npm init -y
    ```

3. Create an `src` folder in `/gatsby-theme-sample-page`.

4. Create a `pages` folder in your `src` folder.

5. Create a `sample.js` file inside your new `pages` folder and add the following code:

    ```
    import React from "react";

    const Sample = () => {
      return (
        <div>
          <h1>Sample page</h1>
        </div>
      );
    };
    export default Sample;
    ```

    This page is very basic and just renders a heading on the page.

6. Navigate to your `gatsby-config.js` file at the root of your project and add the following code:

    ```
    module.exports = {
      // rest of config
      plugins: [
        'gatsby-theme-sample-page',
      // other plugins
      ]
    }
    ```

7. Start your Gatsby development server and navigate to /sample; you should see your sample page.

You may have noticed that a plugin consists of the same building blocks as your Gatsby site. This is one of the reasons why creating plugins is so straightforward in Gatsby. By building a site with this tool, you also inherit the ability to create plugins.

Now that we have seen a basic example, let's try and build something a little more useful and a little more complex. Let's create a plugin that takes a folder of events (in JSON format) and creates a page for each one:

1. First, we're going to need some events that we can source within our plugin. Let's assume each event will have a title, description, location, and date. Create a folder called events within your root directory. Add some JSON files within this folder that are in the following format:

    ```json
    {
        "title": "Elevating Your Hack",
        "description": "Tips & tricks to make your hack
          stand out from the crowd.",
        "location": "King's College",
        "date": "2021-12-25"
    }
    ```

    Ensure that the JSON is valid as errors will cause the plugin to crash out.

2. Create a new folder called gatsby-theme-events-section in your plugins folder.

3. Open a terminal in the gatsby-theme-events-section folder and run the following command:

    ```
    npm init -y
    ```

4. Create an src folder in /gatsby-theme-events-section.

5. Open a terminal in the gatsby-theme-events-section folder and run the following command:

    ```
    npm i gatsby-transformer-json
    ```

    As its name suggests, this installs the transformer plugin for handling JSON.

6. Create a gatsby-config.js file and add the following code:

    ```js
    module.exports = {
      plugins: [
    ```

```
    'gatsby-transformer-json',
    {
      resolve: 'gatsby-source-filesystem',
      options: {
        path: './events',
      },
    },
  ],
};
```

Here, we are adding our newly installed plugin, as well as pointing our plugin to source files from the filesystem that exists in the events directory. These plugins will work together to create a new node for each JSON file within the events directory.

7. Create a gatsby-node.js file and add the following code:

```
const { createFilePath } = require('gatsby-source-
  filesystem');

exports.onCreateNode = ({ node, getNode, actions }) => {
  const { createNodeField } = actions;
  if (node.internal.type === 'EventsJson') {
    const slug = createFilePath({ node, getNode });
    createNodeField({
      node,
      name: 'slug',
      value: slug,
    });
  }
};
```

The onCreateNode function is called whenever a new node is created. Using this function, we can transform nodes by adding, removing, or manipulating their fields. In this specific case, we are adding a slug field if the node is of the EventsJson type. A slug is the address of a specific page on our site, so in the case of our event page, we want every event to have a unique slug where it will render on the site.

8. Prepend your `gatsby-node.js` file with the following code:

```
exports.createPages = async ({ actions, graphql,
 reporter }) => {
  const { createPage } = actions;
  const EventTemplate =
    require.resolve('./src/templates/event');
  const EventsQuery = await graphql('
  {
    allEventsJson {
      nodes {
        fields {
          slug
        }
      }
    }
  }
  ');
  if (EventsQuery.errors) {
    reporter.panicOnBuild('Error while running GraphQL
      query.');
    return;
  }
  const events = EventsQuery.data.allEventsJson.nodes;
  events.forEach(({ fields: { slug } }) => {
    createPage({
      path: 'event${slug}',
      component: EventTemplate,
      context: {
        slug: slug,
      },
    });
  });

};
```

This code should look very familiar as it is very similar to the code we saw in the *Creating templates and programmatic page generation* section of *Chapter 4, Creating Reusable Templates*. Here, we are utilizing the `createPage` function, which allows us to create pages dynamically. Inside this function, we destructure the `actions` object to retrieve the `createPage` function. Then, we tell Gatsby where to find our event template. With these two pieces in place, we are now ready to query our data. You should see a familiar GraphQL query upon selecting the `slug` property from all the events. After this, we can iterate through the events and create a page for each one, providing the `slug` property as context.

9. Create a `templates` folder in `/gatsby-theme-events-section/src`.

10. Create an `event.js` file in `/gatsby-theme-events-section/src/templates` and add the following code:

```
import React from "react";
import { graphql } from "gatsby";

export default function Event({ data }){
  const {
    event: { description, title, location, date },
  } = data;

  return (
    <div className="prose max-w-5xl">
      <h1>{title}</h1>
      <p>
        {date} - {location}
      </p>
      <p>{description}</p>
    </div>
  );
}
```

Here, we take `title`, `location`, `description`, and `date`, which we will retrieve in the page query, and render them on the screen.

11. Append the `events.js` file with the following code:

```
export const pageQuery = graphql'
  query($slug: String!) {
```

```
    event: eventsJson(fields: { slug: { eq: $slug } }) {
      description
      title
      location
      date(formatString: "dddd Do MMMM yyyy")
    }
  }
';
```

Here, we are using `slug` from the context to find the event where `slug` matches in the node's fields. We query for all the data that we need to populate this page by retrieving `title`, `location`, `description`, and `date`, which have been formatted. This is then passed into the template via the `data` prop.

12. Now, let's create a page with all the events listed. Create a `pages` folder in `/gatsby-theme-events-section/src`.

13. Create an `events.js` file in `/gatsby-theme-events-section/src/pages` and add the following code:

```
import React from "react";
import { graphql, Link } from "gatsby";

const Events = ({ data }) => {
  const events = data.allEventsJson.nodes;
  return (
    <div className="prose max-w-5xl">
      <h1>Upcoming Events:</h1>
      {events.map((({ title, location, date, fields: {
      slug } }) => (
        <Link to={'/event${slug}'}>
          <h2>{title}</h2>
          <p>
            {date} - {location}
          </p>
        </Link>
      ))}
    </div>
  );
```

```
};
export default Events
```

Here, we are mapping through our events and creating a `Link` to an event's dedicated page with `title`, `data`, and `location`.

14. Append `events.js` with the following code:

```
export const query = graphql'
  {
    allEventsJson {
      nodes {
        location
        title
        date
        fields {
          slug
        }
      }
    }
  }
';
```

This query will retrieve all the events and return them in a nodes array, which can be retrieved via the `data` prop on the page.

15. You're all done – run your development server and navigate to `localhost:8000/events`. You should see the following output:

Upcoming Events:

Elevating Your Hack

2021-12-25 - King's College

Introduction to Web Development

2021-10-31 - Imperial College

Figure 10.3 – Events page preview

Clicking on an event should take you to its dedicated page:

Elevating Your Hack

Saturday 25th December 2021 - King's College

Tips & tricks to make your hack stand out from the crowd.

Figure 10.4 – Events page preview

You've just made your first local theme plugin. Adding an event to the `events` folder will see it appended to the list and get a dedicated page. If we were to publish this plugin, we could then use it within multiple Gatsby sites to create these pages by simply creating an `events` folder and populating it. No additional configuration is required!

> **Quick Tip**
>
> You'll notice a lack of styling in the examples set out in this chapter. This chapter focuses on the Gatsby APIs that are being utilized and less on styling. By now, you should feel confident enough to create styling for these pages.

Now that we understand how to create both types of plugins, let's learn how we can publish them and contribute them back to the community.

Contributing to the plugin ecosystem

So, you've built a plugin and now you want to use it in a separate Gatsby project? Or perhaps you think the plugin could help other developers? In either case, you'll need to publish your plugin. By publishing your plugin with npm, your plugin will automatically become visible on Gatsby's site plugins page (`https://www.gatsbyjs.com/plugins`). Let's start this journey by looking at a pre-publish checklist.

Pre-publish checklist

Before we publish our plugin, it's important to ensure that we are ready to do so. The following is a suggested pre-publish checklist:

- Ensure your plugin's name explains what it does. This might seem a little trivial but naming your plugin in a way that makes it clear what it does will make it easier to find online.

- Ensure your plugin's name is unique. Two npm packages cannot share the same name, so you mustn't try and deploy a package with a name that is already in use. To check whether your name is in use, visit `https://www.npmjs.com/` and search for your plugin's name.

- Ensure your plugin adheres to the naming convention that was outlined in the *Understanding Gatsby plugins* section. This is the way that Gatsby determines which npm packages are Gatsby plugins so that it can add them to their site.

- Ensure your plugin has a comprehensive `README.md` file. This file will be picked up by Gatsby and included within the plugin ecosystem, so it's vitally important that the `README.md` file explains what your plugin does and how to use it. You should include the specific configuration options that might be required.

- Check that both React and Gatsby are peer dependencies.

- Ensure your code has been tested properly. Unit tests are so important, but even more so if you're about to pass your code onto others. Aim for 100% test code coverage.

> **Important Note**
> If you ended up changing your plugin's name, be sure that this new name is reflected in the `package.json` file, as well as the folder's name. Having the old name anywhere can be confusing when you're using/searching for your plugin later down the line.

Now that we have gone through our checklist, let's learn how to publish a plugin.

Publishing a plugin

Publishing a Gatsby plugin follows the same process as publishing any npm package. Let's learn how to do this:

1. Ensure you have an npm account. If you do not, you can create one at `https://www.npmjs.com/signup`.

2. Log into the npm CLI from your terminal by running the following command:

```
npm login
```

The CLI will ask for your name, email, and password as part of the login process.

3. Navigate your terminal to your Gatsby plugin's directory. *This is vitally important.* If you continue these steps within your root directory, you will end up accidentally releasing your entire site as a package, so ensure you have navigated to the plugin.

4. Finally, run the `publish` command:

```
npm publish
```

If this command finishes successfully, congratulations! Your plugin is now available via npm. If, however, you saw permission errors, this is most likely because your plugin name is not unique. Refer back to the *Pre-publish checklist* section and if the name is clashing, change the name in the package.json file and retry.

> **Quick Tip**
>
> After your first publish, you will most likely find things you want to change. If you follow these instructions again, be sure to bump the version number in your package.json file as npm will reject a publish with the same version number.

Now that your plugin has been published, it should be visible on the Gatsby website plugins page within 24 hours.

Summary

In this chapter, we learned what a Gatsby plugin is and what types exist. We learned about local plugin development and how to create source and theme plugins. We created both source and theme plugins and then tested them locally by including them on our site. We then learned about sharing plugins online. We discussed what you should consider before deploying a plugin and then learned how to share a plugin by publishing them online via npm. By completing this chapter, you should now feel confident that you can create and share source and theme plugins with ease. This has been a brief introduction to a massive topic, and I hope you can build on this knowledge to create plugins for any use case.

In the next chapter, we will look at another advanced concept – authentication. We will learn how to create login experiences on your website.

11
Creating Authenticated Experiences

Within the context of this book, **authentication** is the act of verifying that a user is who they say they are within a website. Once their identity has been verified, we can show the individual content that's only meant for them. This might be their profile page, delivery address, bank details, and more. In this chapter, we're going to focus more on how to implement routing for use with authentication services instead of focusing on how to implement authentication services or what content to display when a user is authenticated. We will remind ourselves of how this is done in traditional React applications before applying this knowledge to Gatsby sites with two different client-side implementations.

In this chapter, we will cover the following topics:

- Routing and authentication in React applications
- Authentication using client-only routes within Gatsby
- Site-wide authentication using context within Gatsby

Technical requirements

To complete this chapter, you will need to have completed *Chapter 10, Creating Gatsby Plugins*. You will also need a GitHub account.

The code for this chapter can be found at `https://github.com/ PacktPublishing/Elevating-React-Web-Development-with-Gatsby-4/ tree/main/Chapter11`.

Routing and authentication in React applications

To achieve authenticated experiences, we will be using **routing**. Before jumping into how we do this in Gatsby, let's familiarize ourselves with how routing works within React applications. Routing is the process of navigating a user around different parts of an application.

For this example, I will be bootstrapping a React project using `create-react-app`. I have included steps for its installation but feel free to skip them and use your own React implementation. *Keep this section's demo separate from your Gatsby project.*

> **Important Note**
> In the following example, we will be using the `@reach/router` package for routing. Gatsby uses `@reach/router` under the hood, so by using the package here in React, it will be easy to recognize patterns when we move on to implementing them in Gatsby.

As React developers, routing is a common part of building applications – let's remind ourselves of the routing basics:

1. Create a new folder for this demo. Open a terminal within this new folder and run the following command:

   ```
   npx create-react-app .
   ```

2. In the same terminal, run the following command:

   ```
   npm i @reach/router
   ```

   This will install the `@reach/router` package within the project.

3. Open `src/App.js` and replace it with the following code:

```
import { Router, Link } from "@reach/router";

const Nav = () => (
  <nav>
    <Link to="/">Homepage</Link> | <Link
      to="about">About Me</Link>
  </nav>
);

{/* Code continued in next step */}
```

Here, we have imported `Router` and `Link` from the `@reach/router` package. We have also created a `Nav` component that we can use to access the routes. This `Nav` component utilizes the `Link` component from `@reach/router` to provide navigation between routes.

4. Append `src/App.js` with the following code:

```
const HomePage = () => (
  <div>
    <Nav />
    <h1>Homepage</h1>
  </div>
);

const AboutPage = () => (
  <div>
    <Nav />
    <h1>About Me</h1>
  </div>
);
{/* Code continued in next step */}
```

Here, we have defined a couple of dummy components to route between – a home page and an `about` page. This should all be very familiar to you.

5. Finally, append `src/App.js` with the following code:

```
function App() {
  return (
    <Router>
      <HomePage path="/" />
      <AboutPage path="about" />
    </Router>
  );
}

export default App;
```

This is where the magic happens. By wrapping the components in a `Router` component, we can switch out which component is displayed based on the current URL path. In this instance, if the user is at the / path (the route URL) they will see the `HomePage` component, while if `path` is `/about`, they will see the `AboutPage` component. They can use the `Nav` component within these two pages to navigate between the two of them.

6. Start the project by running `npm start` from the root directory to try it out.

It's important to remember that navigating between routes is fast because all the routes are loaded when the router renders. As we move into Gatsby, it's important to make sure we only use routers when it is necessary as we might be adding page weight to include components that a user may never have any intention of seeing.

Now that we have gone through a basic routing example, let's start adding pages that can only be accessed once a user has logged in. We will do this with **private routes**.

Private routes

A private route behaves the same as the other components that are wrapped in a `Router`, except it has an authentication condition. If the condition is not satisfied, instead of seeing the requested content, the user will be redirected to a login screen to authenticate. Let's try this out now by turning our about page that was previously public into a private route:

1. First, we are going to need to define our authentication condition. For this example, we are going to keep it simple. To be considered "authenticated," the user must have called the `login` function, which we will trigger via a button on the login page. To achieve this condition, we are going to create a context that can store the current authentication state. Create a new file called `auth-context.js` and add the following code:

    ```
    import React, { useState, useContext } from "react";
    import { navigate } from "@reach/router";
    const AuthContext = React.createContext();

    export const AuthProvider = ({ ...props }) => {
      {/* Code continued in next step */}
    };
    export const useAuth = () => useContext(AuthContext);
    export default AuthContext;
    ```

    Here, we are setting up the boilerplate of our authorization context. We are creating a `useAuth` hook to access the context values that we will be defining in the next step.

2. Within the `auth-context.js` file's `AuthProvider`, add the following code:

    ```
    Const [authenticated, setAuthenticated] =
      useState(false);
      const login = () => {
        // Make authentication request here and only
          trigger the following if successful.
        setAuthenticated(true);
        navigate("/");
      };
      const logout = () => {
        setAuthenticated(false);
        navigate("/login");
    ```

```
  };
  return (
    <AuthContext.Provider
      value={{
        login,
        logout,
        authenticated,
      }}
      {...props}
    />
  );
```

Here, we have created a useState value called authenticated to track whether the user is authenticated or not. We then created a login function that sets authenticated to true. It is within this function that you would make a request to your authentication service and verify the user before authenticating. Most likely, you will also have some information about the user that you could store in your state or local storage. If you do add additional information, be sure to clear it within the logout function. For the time being, the logout function just sets authenticated to false and navigates a user back to the login page. Within AuthContext.Provider, we expose the login and logout functions, as well as the authenticated state, to the rest of the application.

3. Navigate to your React application's index.js file and modify it with the following code:

```
import React from "react";
import ReactDOM from "react-dom";
import "./index.css";
import App from "./App";
import { AuthProvider } from "./auth-context";
import reportWebVitals from "./reportWebVitals";

ReactDOM.render(
  <React.StrictMode>
    <AuthProvider>
      <App />
    </AuthProvider>
  </React.StrictMode>,
```

```
        document.getElementById("root")
    );
```

Without wrapping our application in `AuthProvider`, we would not be able to access the authentication context within the application.

4. Our authentication condition is now defined, so we can utilize it to create a private route component. Create a new file called `PrivateRoute.js` and add the following code:

```
import React from "react";
import { navigate } from "@reach/router"
import { useAuth } from "./auth-context";

const PrivateRoute = ({
  component: Component,
  ...rest
}) => {
  const { authenticated } = useAuth();
  if (!authenticated) {
    navigate("/login");
    return null;
  }
  return <Component {...rest} />;
};
export default PrivateRoute;
```

This `PrivateRoute` component uses the `authenticated` state from the `useAuth` hook to conditionally render a given component. If the user is authenticated, the component will be rendered. If, however, they are not authenticated, the user will be navigated to the `login` route instead.

5. Return to your `App.js` file and update the file with the following imports:

```
import { useAuth } from "./auth-context";
import PrivateRoute from "./PrivateRoute";
```

Here, we are importing the `useAuth` hook and our newly created `PrivateRoute` component.

6. Modify the `HomePage` component with the following code:

```
import { Router, Link } from "@reach/router";

// Predefined Nav Component Here.

Const HomePage = () => {
  const { authenticated } = useAuth();
  return (
    <div>
      <Nav />
      <h1>You are {authenticated ? "logged in" :
        "logged out"}.</h1>
    </div>
  );
};
```

We can use the `authenticated` state from the `useAuth` hook to give the user some indication of their authentication status.

7. Modify the `AboutPage` component with the following code:

```
const AboutPage = () => {
  const { logout } = useAuth();
  return (
    <div>
      <Nav />
      <h1>About Me</h1>
      <button onClick={logout}>Logout</button>
    </div>
  );
};
```

This is the path we intend to make private in this demo. If a user is on this page, we can assume they have been authenticated and render a logout button to allow them to trigger the `logout` function.

8. Add a `LoginPage` component to `App.js`:

```
const LoginPage = () => {
  const { login } = useAuth();
  return (
    <div>
      <Nav />
      <h1>Login Page</h1>
      <button onClick={login}>Login</button>
    </div>
  );
};
```

This is a basic implementation that uses the `login` function from `useAuth` to log a user in. In your application, you would probably want to flesh this out with inputs where users enter their email and password. You would then pass this to the `login` function so that it can be used as part of the authorization request.

9. Finally, update your `App` function so that it includes your `PrivateRoute` component:

```
function App() {
  return (
    <Router>
      <HomePage path="/" />
      <LoginPage path="login" />
      <PrivateRoute component={AboutPage} path="about" />
    </Router>
  );
}

export default App;
```

10. Start the project by running `npm start` from the root directory. If you try to navigate to `/about`, you will notice that you will be redirected to `/login` until you have clicked the login button.

We now have a firm grasp on routing and private routes, so let's take the knowledge we have gained in this section and apply it to Gatsby.

Authentication using client-only routes within Gatsby

While not common practice, we can use routers inside Gatsby pages too. Normally, Gatsby abstracts all the routing away so that we don't have to worry about it, but authentication is one example where we need to bring the control over routing back into our hands. We will be creating what is known as client-only routes. To demonstrate this within our project, we are going to create a page at `/private`. As its name might suggest, this path contains a private page that we will lock behind authentication. Let's get started:

> **Important Note**
> This example will conflict with the *Site-wide authentication using context within Gatsby* section's code. It's best to choose one of these two methods to implement instead of trying to combine them.

1. Create a new folder inside `src` called `context`.

2. Create a new file called `auth-context.js` and add the following code:

```
import React, { useState, useContext } from "react";
import { navigate } from "@reach/router";
const AuthContext = React.createContext();

export const AuthProvider = ({ ...props }) => {
  {/* Code continued in next step */}
};

export const useAuth = () => useContext(AuthContext);
```

Here, we are setting up an authentication context in the same way we did within the example code from the *Routing and authentication in React applications* section. Note that we are still importing `navigate` from `@reach/router` instead of the Gatsby library.

3. Add the following within `AuthProvider`:

```
const [authenticated, setAuthenticated] =
  useState(false);
const login = async () => {
  // Make authentication request here and only
    trigger the following if successful.
  setAuthenticated(true);
```

```
        navigate("/private")
    };
    const logout = () => {
        setAuthenticated(false);
    };
    return (
        <AuthContext.Provider
            value={{
                login,
                logout,
                authenticated,
            }}
            {...props}
        />
    );
```

We set up this `auth-context.js` file in the same way we did with the React demo, except this time, we navigate to `/private` on a successful login.

> **Important Note**
>
> Within this section, you will see code that looks very similar to the React demo from the previous section. Please note that while they are similar, they are not the same. Don't be tempted to copy and paste them from the React example.

4. Add the following to your `gatsby-browser.js` and `gatsby-ssr.js` files:

```
import React from "react";
import { AuthProvider } from "./src/context/auth-
    context";

export const wrapPageElement = ({ element }) => {
    return <AuthProvider>{element}</AuthProvider>;
};
```

We want to ensure that the authentication context is available throughout the application. By adding the preceding code to both `gatsby-browser.js` and `gatsby-ssr.js`, we can be sure it is accessible everywhere.

5. Create a new file within `src/components` called `PrivateRoute.js`.

6. Add the following code to the newly created `PrivateRoute.js`:

```
import React from "react";
import { navigate } from "gatsby";
import { useAuth } from "../context/auth-context";

const PrivateRoute = ({
  component: Component,
  location,
  basepath,
  ...rest
}) => {
  const { authenticated } = useAuth();
  if (!authenticated) {
    navigate(basepath + "/login");
    return null;
  }
  return <Component {...rest} />;
};
export default PrivateRoute;
```

This is a Gatsby-friendly implementation of the `PrivateRoute` component. Note that we are switching out the `@reach/router` part of `navigate` for Gatsby's implementation. This is because Gatsby's implementation will handle the redirect in a way that is suitable for a Gatsby project. Without this switch, you will be presented with a white screen when `navigate` is called. You will also notice that we are passing in a prop called `basepath`. As our router will not sit at the top of the application, the `PrivateRoute` component must know the router's base path location to ensure it navigates the respective users to it.

7. Create a new folder inside `src/pages` called `private`.

8. Inside this new folder, create a new file called `[...].js`. Using Gatsby's **file-system-route API**, this format creates a wildcard route that matches anything whose path begins with `/private`. This step is vitally important as Gatsby does not know the router we will set up, so it needs to understand that if it sees a path beginning with `/private`, such as `/private/login`, it needs to be handled by this file instead of erroring out with a 404 status code.

9. Add the following code to `src/pages/private/[...].js`:

```
import React from "react";
import { Router } from "@reach/router";
import Layout from "../components/layout/Layout";
import PrivateRoute from "../components/PrivateRoute";
import { useAuth } from "../context/auth-context";

const LoginPage = () => {
  const { login } = useAuth();
  return (
    <Layout>
      <h1>Login Page</h1>
      <button onClick={login}>Login</button>
    </Layout>
  );
};

const AuthenticatedPage = () => {
  const { logout } = useAuth();
  return (
    <Layout>
      <h1>Authenticated Page</h1>
      <button onClick={logout}>Logout</button>
    </Layout>
  );
};
```

Here, we are defining the two possible paths that will be visible. Either you will be shown `AuthenticatedPage` or, if you are not logged in, you will see the login page. These components both make use of the `useAuth` hook to retrieve the functions they require.

10. Append the following code to `src/pages/private/[...].js`:

```
function PageWithRouter() {
  const basepath = "/private";
  return (
    <Router basepath={basepath}>
```

```
        <LoginPage path="login" />
        <PrivateRoute
          basepath={basepath}
          component={AuthenticatedPage}
          path="/"
        />
      </Router>
    );
  }


export default PageWithRouter;
```

Within this step, we have defined our `basepath` – this must match the Gatsby page's path (which, in this instance, is `/private`). We pass this value as a prop both to `Router` and `PrivateRoute`. This example is different from the React example in that the base path is the path that requires authentication.

11. Start the project by running `npm start` from the root directory. If you try to navigate to `/private`, you will notice that you are redirected to `/private/login`, and clicking the login button will redirect you to `/private`.

With that, we've learned how to add routing within a particular section of our Gatsby site. Now, let's turn our attention to an implementation that you can use when your whole site requires authentication.

Site-wide authentication using context within Gatsby

There may be situations where you want the entirety of your Gatsby site to be behind authentication. For example, you may have made a documentation site only meant for employees of your company. Let's look at how we can use context to turn every page into a private route:

1. First, let's create a login component in the `components` folder. Call this file `Login.js` and add the following code to it:

```
import React from "react";

const Login = ({login}) => {
  return <button onClick={login}>Login</button>;
```

```
};
```

```
export default Login;
```

You'll notice that, unlike the last `Login` component we created, we are not retrieving the `login` function from the context. The reason for this will become clear when we create the context.

2. Create a folder called `context` in `src`.

3. Create a file in `context` called `auth-context.js` and add the following code:

```
import React, { useState, useContext } from "react";
import Login from "../components/Login";
const AuthContext = React.createContext();

export const AuthProvider = ({ ...props }) => {
{/* Code continued in next step */}
};

export const useAuth = () => useContext(AuthContext);

export default AuthContext;
```

Here, we are setting up the authentication context in the same way we did in the *Routing and authentication in React applications* section but with one addition. We are now also importing our `Login` component into our authentication context.

4. Add the following code within `AuthProvider`:

```
const [authenticated, setAuthenticated] =
  useState(false);

const login = async () => {
  // Make authentication request here and only
    trigger the following if successful.
  setAuthenticated(true);
};
const logout = () => {
  setAuthenticated(false);
};
```

```
if (!authenticated) {
  return <Login login={login} />;
}
return (
  <AuthContext.Provider
    value={{
      login,
      logout,
      authenticated,
    }}
    {...props}
  />
);
```

If the user is not authenticated, the provider will return the Login component, being sure to pass in the login function as a prop. This does not cause a route change, which can be a great benefit. When a user navigates to a page, their requested path is not lost by navigating away to a login page and, as such, when the user has successfully logged in, they will jump right back into the application in the place they intended to be. As a developer, this can stop you from having to pass redirect URLs around in the browser, which can be a hassle.

If, for some reason, you want to keep a few pages public, you can check for the path in this conditional statement and allow some paths to be accessible, even without being authenticated. Note that even on these pages, the Login component will be loaded in, despite the fact it is not being used and will add unnecessary page weight.

5. Add the following to your gatsby-browser.js and gatsby-ssr.js files:

```
import React from "react";
import { AuthProvider } from "./src/context/auth-
  context";

export const wrapPageElement = ({ element }) => {
  return <AuthProvider>{element}</AuthProvider>;
};
```

We want to ensure that the authentication context is available throughout the application. By adding this file to both `gatsby-browser.js` and `gatsby-ssr.js`, we can ensure it is accessible everywhere.

6. Start the project by running `npm start` from the root directory and navigate to any page on the site. You should find that you are prompted to log in before being able to view the page.

Now that we have looked at two different ways to achieve authenticated experiences within Gatsby applications, let's summarize what we have learned.

Summary

In this chapter, we explored routing and authenticated experiences. We reminded ourselves of how routing works in React and created private routes for use with `@reach/router`. Then, we ported this knowledge into Gatsby and created a private page that was only accessible by logging in. Finally, we investigated how we can use context to wrap our whole application in authentication for situations that require it.

In the next chapter, we will learn about another advanced concept – how to use sockets to create experiences that make use of real-time data.

12
Using Real-Time Data

Have you ever ordered food and watched as it made its way closer to your destination without you having to refresh the page? You may have also seen this with package deliveries or ride-hailing apps. All of these make use of **real-time data**. This is a form of data that is presented as soon as it is acquired. So, in these examples, as soon as the service you are using has the food, package, or car's location, it will relay that information to you. The most common way that convenience sites and messaging applications enable real-time data is by using **web sockets**.

In this chapter, we will cover the following topics:

- Introduction to web sockets
- Socket.io in action
- Live site visitor count
- Gaining further insights with rooms

Technical requirements

To complete this chapter, you will need to have completed *Chapter 11, Creating Authenticated Experiences*.

The code for this chapter can be found at `https://github.com/PacktPublishing/Elevating-React-Web-Development-with-Gatsby-4/tree/main/Chapter12`.

Introduction to web sockets

A web socket is a bi-directional communication channel between a client and a server. Unlike **REST** requests, the socket connection's channel remains open for the client and the server to push messages to and from each other whenever they need, instead of closing when a response is received. This kind of communication is commonly associated with low latency, which means it can handle high volumes of data with minimal delay.

So, how does it work? To start, the client sends an HTTP request to a server, asking it to open a connection. If the server agrees, it will send back a response with a status of 101, indicating that it will be switching protocols. At this point, the handshake is complete and a **TCP/IP** connection is left open, allowing messages to pass back and forth between the two devices. This connection will remain open until one of the devices disconnects or loses its connection.

One of the most popular socket implementations in the JavaScript world is *socket.io*, which consists of two parts – a Node.js server and a JavaScript client library. We'll look at socket.io in action by creating a minimal demo in the next section.

> **Quick Tip**
> Note that there are also several other implementations of the *socket.io* server and client libraries available in languages other than JavaScript. This may be helpful if you want to combine a socket server (which we will create in this chapter) with more than just your Gatsby site.

Socket.io in action

In this demo, we will make a server that accepts a socket connection. When it receives a message from the client, it will log it to the console. Let's start by creating the server and then move on to the client:

1. Create a folder called `server` in your root directory.

2. Open a terminal in the `server` folder and run the following command:

```
npm init -y
```

This will set up an empty npm package in the folder.

3. In the same terminal, run the following command:

```
npm i express socket.io
```

Here, we are installing the `express` dependency for creating our server and the `socket.io` library.

4. Create an `app.js` file in the `server` folder and add the following code:

```
const PORT = 3000
const express = require("express");
const server = express()
  .listen(PORT, () => console.log('Listening on
    ${PORT}'));
```

This creates a minimal Express server that listens for requests on port `3000`. As we have already learned, this socket connection is established with an HTTP request, and it requires an HTTP server to do this.

5. Verify that the server is working by opening a terminal within the `server` folder and running the following command:

```
node app.js
```

If the server starts, you will see `Listening on 3000` printed to the console.

6. Update the `app.js` file with the following code:

```
const PORT = 3000
const express = require("express");
const server = express()
  .listen(PORT, () => console.log('Listening on
    ${PORT}'));
const io = require("socket.io")(server);
io.on("connection", (socket) => {
  socket.on("message", (msg) => {
    console.log("message: " + msg);
  });
});
```

Here, we are passing the server instance into `socket.io` for the client-server handshake. We then tell our socket server how to handle events from clients. In this instance, if a client socket sends an event of the `message` type, we log it to the console.

7. Before moving on, we must add a **CORS** policy for our HTTP and socket configuration. Without this, your browser will not be able to access the server as the cross-origin policy will be blocked. To do this, open a terminal within the `server` folder and run the following command:

```
npm i cors
```

This installs the `cors` library, which acts as middleware within our Express application to enable CORS.

8. Now, update your `app.js` file with the following code:

```
const PORT = 3000;
const express = require("express");
var cors = require("cors");
var allowlist = ["http://localhost:8000"];
var corsOptions = {
  origin: function (origin, callback) {
    var originIsAllowlisted =
      allowlist.indexOf(origin) !== -1;
    callback(null, originIsAllowlisted);
  },
};
const server = express()
  .use(cors(corsOptions))
  .listen(PORT, () => console.log('Listening on
    ${PORT}'));

const io = require("socket.io")(server, {
  cors: {
    origin: corsOptions.origin,
  },
});
io.on("connection", (socket) => {
  socket.on("message", (msg) => {
```

```
        console.log("message: " + msg);
    });
});
```

This CORS setup uses an `allowlist` of origins that are allowed to access the server. The middleware checks the origin of any request to ensure that the origin is present in this list and is therefore allowed. If a request comes from an origin that is not on the list, the cross-origin request will be blocked. In this case, we have added `localhost:8000`, which is the default development port for Gatsby. If this changes or you are hosting the application, this list will need to be updated.

9. Now that we have set up our socket server, let's interact with it from our Gatsby site by using Gatsby as the socket client. Navigate back to the root of your Gatsby site. Open a terminal here and run the following command:

    npm i socket.io-client

    As the name of the library might suggest, this installs the socket.io client library that we will be using to communicate with our web socket server.

10. Create a new file within your `pages` folder called `socket.js` and add the following code to it:

```
import React from "react";
import openSocket from "socket.io-client";
import Layout from "../components/layout/Layout";
export default function SocketDemo() {
  const [socket, setSocket] = React.useState(null);
  const [value, setValue] = React.useState("");
  React.useEffect(() => {
    const newSocket =
      openSocket("http://localhost:3000");
    setSocket(newSocket);
    return () => newSocket.close();
  }, [setSocket]);
```

The standard setup for this page is the same as any other Gatsby page. We have additionally imported our new `socket.io-client` package. Inside a `useEffect`, we create the socket connection by using the default export from `socket.io-client` with the server URL string as an argument. In our case, the server port was defined as `3000`, so we added `http://localhost:3000`. This one line of code abstracts all the logic around the client-server handshake, so all you need to focus on is firing the messages you want to send. We then set the socket in our `useState` so that we can use it within the page. It's best to create the socket connection in `useEffect` as we only want this connection to be established once. If the page re-renders, we do not want the socket to reconnect as this would be perceived as a new connection by the server. The `return` statement in our `useEffect` ensures that the socket connection is closed when the component dismounts.

11. Continue editing `socket.js` and add the following code:

```
const sendMessage = () => {
  socket && socket.emit("message", value);
};
return (
  <Layout>
    <div className="max-w-5xl mx-auto py-16 lg:py-24
      flex flex-col prose space-y-2 ">
      <h1>Message The Server</h1>
      <label htmlFor="message">Your Message:</label>
      <input
        id="message"
        className="border-blue-700 border-2"
        onChange={(e) => setValue(e.target.value)}
      />
      <button onClick={sendMessage} className="btn">
        Send message
      </button>
    </div>
  </Layout>
);
}
```

We've set up a simple form here. The input updates the value of the state, which we can then send to the server by clicking the **Send message** button. Upon clicking this button, the `sendMessage` function is called, which uses `socket.emit` (if there is a socket available in the state), which emits a message from this client to the server. The first argument is the message type, while the second argument is the body of the message. In this case, we are just sending a string, but you could also send an object with multiple key-value pairs. If you were to send an object, there is no need to `JSON.stringify` it as the library handles all that for you.

12. Start your Gatsby development server and ensure your socket server is also running. Navigate to `localhost:8000/socket`, type in a message, and click **Send message**. With any luck, the contents of your message should now be logged within your server's terminal. Congratulations – you've just sent your first message via sockets!

Now, let's expand this demo so that the client can receive communication back from the server. As an example, let's make the server return one of three random greetings when it receives a message:

1. First, we need to modify how our server handles messages. Modify the socket server's connection configuration with the following code:

```
io.on("connection", (socket) => {
  socket.on("message", (msg) => {
    console.log("message: " + msg);
    socket.emit(
      "message",
      ["Hi there!", "Hello!",
      "Howdy"][Math.floor(Math.random() * 3)]
    );
  });
});
```

Now, as well as logging the messages that are received from a client, we emit something back to that same client. In this case, we are choosing a random greeting to send back.

2. With our Gatsby page, we need to tell it to expect and handle messages of a certain type. This works like event listeners, so this should feel familiar to you:

```
export default function SocketDemo() {
  const [socket, setSocket] = React.useState(null);
  const [value, setValue] = React.useState("");
  const [serverMessages, setServerMessages] =
  React.useState([]);
  React.useEffect(() => {
    const newSocket =
      openSocket("http://localhost:3000");
    setSocket(newSocket);
    return () => newSocket.close();
  }, [setSocket]);
  React.useEffect(() => {
    if (socket) {
      socket.on("message", (message) => {
        setServerMessages((currentMessages) =>
        [...currentMessages, message]);
      });
    }
  }, [socket, setServerMessages]);
  const sendMessage = () => {
    socket && socket.emit("message", value);
  };

  // render in next step
```

Here, we created a new useState hook to store the server messages. As we may receive more than one, we set this to an empty array that we can push elements to. Then, we defined a second useEffect. If the socket connection has been established, this function listens for messages from the server of the message type. If it receives one, it adds the body of the message to the server message list.

3. Update the render of the page component:

```
return (
  <Layout>
    <div className="max-w-5xl mx-auto py-16
      lg:py-24 flex flex-col prose space-y-2 ">
      <h1>Message The Server</h1>
      <label htmlFor="message">Your Message:</label>
      <input
        id="message"
        className="border-blue-700 border-2"
        onChange={(e) => setValue(e.target.value)}
      />
      <button onClick={sendMessage} className="btn">
        Send message
      </button>
      <label>Server Messages:</label>
      <ul>
        {serverMessages.map((message, index) => (
          <li key={index}>{message}</li>
        ))}
      </ul>
    </div>
  </Layout>
);
}
```

Within the render, we can map through the server messages and render them to the screen in a bulleted list.

4. Start your Gatsby development server and ensure your socket server is also running. Navigate to `localhost:8000/socket`, type in a message, and click **Send message**:

Figure 12.1 – Socket demonstration page

Your message should be logged within the server's terminal, but additionally, the server should have also sent a message back. It should be visible underneath the **Send Message** button. The speed at which this happens can feel crazy. And when the connection is good, it can almost feel like the server message is being triggered by your button press.

We now have a clear understanding of how socket connections work and we have managed to send messages between the client and the server. Now, let's apply what we have learned and build something useful for our Gatsby site with this technology – a live visitor count in our site footer.

Live site visitor count

The setup for this will need to be a little different from the previous example since, in the *Socket.io in action* section, the socket connection was isolated to a single page. However, our site footer is not on a single page but every page! An implementation of this that would work well is wrapping the site in some context. By doing this, we would be able to access the count in other parts of the application if we needed to. Let's try this approach together:

1. Modify the socket server's connection configuration with the following code:

```
io.on("connection", (socket) => {
  io.emit("count", io.engine.clientsCount);
  socket.on("disconnect", function () {
    io.emit("count", io.engine.clientsCount);
  });
});
```

   We've changed this configuration quite a bit, so let's break it down. When a new socket connects to the server, we use `io.emit`. This function sends a message to all the connected clients instead of a single socket. The socket type is `count` and the body contains `io.engine.clientsCount`, which is a count of the number of connected clients. If you use this whenever a new client connects, everyone will know that the count has changed. Then, we have to make sure that the count for clients is updated on disconnect too. For that, we trigger the same `io.emit` when the server has seen a client drop off.

2. Create a new folder inside `src` called `context` if you don't already have one.

3. Create a new file called `stats-context.js` and add the following code:

```
import React, { useState, useContext } from "react";
import openSocket from "socket.io-client";

const socket = openSocket("http://localhost:3000");
const StatsContext = React.createContext();

export const StatsProvider = ({ ...props }) => {
  {/* Code continued in next step */}
};
export const useStats = () =>
```

```
  useContext(StatsContext);
export default StatsContext;
```

Here, we are setting up the boilerplate of our stat's context. We create a `useStats` hook to access the context values that we will be defining in the next step.

4. Add the following code within `StatsProvider`:

```
const [socket, setSocket] = React.useState(null);
const [liveVisitorCount, setLiveVisitorCount] =
  useState(0);
React.useEffect(() => {
  const newSocket =
    openSocket("http://localhost:3000");
  setSocket(newSocket);
  return () => newSocket.close();
}, [setSocket]);
React.useEffect(() => {
  if (socket) {
    socket.on("count", (count) => {
      setLiveVisitorCount(count);
    });
  }
}, [socket, setLiveVisitorCount]);
return (
  <StatsContext.Provider
    value={{
      liveVisitorCount,
      connected: socket && socket.connected,
    }}
    {...props}
  />
);
```

Within the page level demo, we set up the socket using a `useEffect`. We do the same thing here to ensure it only happens one time. Then, we create a second `useEffect` that, when connected to the server, will listen for messages of the `count` type. If one is received, it updates the count in state, which will then be available throughout the application via the `useStats` hook.

5. Update your `gatsby-browser.js` and `gatbsy-ssr.js` files with the following code:

```
import React from "react";
import { StatsProvider } from "./src/context/stats-
   context";
export const wrapPageElement = ({ element }) => {
  return <StatsProvider>{element}</StatsProvider>;
};
```

We want to ensure that the count's context is available throughout the application. By adding this file to both the `gatsby-browser.js` and `gatsby-ssr.js` files, we can be sure it is accessible everywhere.

6. Create a `VisitorCountBadge.js` file in `src/components/layout` and add the following code to it:

```
import React from "react";
import { useStats } from "../../context/stats-
   context";
const VisitorCountBadge = () => {
  const { liveVisitorCount, connected } = useStats();
  return (
    <p className={'${connected? "bg-blue-200" :"bg-
      red-200"} px-2 py-1 inline-block rounded'}>
      Visitors: {liveVisitorCount}
    </p>
  );
};
export default VisitorCountBadge;
```

Here, we are making use of the `useStats` hook to retrieve `liveVistorCount` and the connected status. The color of the badge is dependent on the connection's status – if it is blue, then we are connected to the server; if not, it will be red. Then, we render `liveVistorCount` within this badge so that it is visible to the user.

Important Note

Here, we are using colors to signify the application state as an example only. Color alone should never be used to signify application state in production as it can leave your application inaccessible to colorblind users. It is better to combine color with another visual indicator, such as text, or at the very least an `aria-label`.

7. Update your `Footer` component file with the following code:

```
import React from "react";
import VisitorCountBadge from "./VisitorCountBadge";
const Footer = () => (
  <footer className="px-2 border-t w-full max-w-5xl
    mx-auto py-4">
    <div className="flex justify-between w-full">
      <VisitorCountBadge />
      <div className="flex items-center">
        <p>Your name here.</p>
      </div>
    </div>
  </footer>
);
export default Footer;
```

Where you want to use the badge and how you style it is entirely up to you. But by adding it to the `Footer` component, it will be visible on every page that utilizes our `Layout` component.

8. Start your Gatsby development server and ensure your socket server is also running. Navigate to `localhost:8000` and you should see the visitor count. If you duplicate your browser tab, the visitor count will rise, while if you close a tab, the count will fall. Finally, if you close the terminal with the socket server running, you should see the badge change to red, indicating it has lost connection to the server.

We have now implemented a working current visitor count. Let's build on this feature by using rooms.

Gaining further insights with rooms

There is one element to socket.io events that we have not talked about yet but could be of great benefit in our application – **rooms**. Rooms are channels that a socket can join and leave. The server can emit messages to a room to broadcast an event to a subset of the clients connected to the server.

To demonstrate the concept of rooms, we will be breaking down our visitor count into more granular stats. Not only will we display to the user the count of total users on the site, but we will also provide them with the details of how many people are on their current page of the site. Let's get started:

1. Update your server/app.js file's socket code so that it includes a new event:

```
// defined at top of file
const pathToRoom = (path) => 'Page-${path}';
// defined in socket configuration
socket.on("page-update", ({ currentPage, previousPage
  }) => {
    if (previousPage) {
      const previousRoom = pathToRoom(previousPage);
      socket.leave(previousRoom);
      io.to(previousRoom).emit(
        "page-count",
        io.sockets.adapter.rooms.get(previousRoom)?.size
      );
    }
    const roomToJoin = pathToRoom(currentPage);
    socket.join(roomToJoin);
    io.to(roomToJoin).emit(
      "page-count",
      io.sockets.adapter.rooms.get(roomToJoin).size
    );
  });
```

We now expect clients to send us a new event of the page-update type. The body contains a currentPage and an optional previousPage for the client. We will use these two pieces of information to make them join the room for their current page and remove them from the room for their previous page.

We have defined a function called `pathToRoom` that we use to take the path where the user is and turn it into a string that we can use as a room identifier. If the client has sent a previous page, we know that this is not the first page on the site they have visited, so they need to be removed from the `previousPage` room. To do this, we can call the `socket.leave` function with the room identifier as the argument. We can then use `io.to(previousRoom).emit` to emit the new reduced count to users still on that page. After that, we can use `currentPage` to determine the new room that the user should join and emit the new count to users in that room (including the new user).

> **Quick Tip**
>
> `socket.leave` and `socket.join` are server-side only. Sockets cannot leave and join rooms on the client site.

2. Update the `disconnect` event with the following code:

```
socket.on("disconnect", function () {
    io.emit("count", io.engine.clientsCount);
    for(room of io.sockets.adapter.rooms){
      io.to(room[0]).emit(
        "page-count",
        io.sockets.adapter.rooms.get(room[0])?.size
      );
    };
});
```

When a socket disconnects, we loop through all open rooms and emit the new number of clients to each of them.

3. Update your `gatsby-browser.js` and `gatbsy-ssr.js` files with the following code:

```
import React from "react";
import { StatsProvider } from "./src/context/stats-
  context";
export const wrapPageElement = ({ element, props }) => {
  return <StatsProvider
  location={props.location}>{element}</StatsProvider>;
};
```

Here, we are passing in the `location` object that Gatsby provides via props to `StatsProvider`. The `location` object contains a `pathname` variable, which will tell us what path the user is currently at.

4. Navigate to your `stats-context.js` file and update the `StatsProvider` arguments:

```
export const StatsProvider = ({ location, ...props })
 => {
 // Code continued in next step
 }
```

We will need to use the location that we are now passing in, so let's de-structure it with props.

5. Add two new React hooks to the top of `StatsProvider`:

```
const [pageVisitorCount, setPageVisitorCount] =
  useState(0);
const previousLocation = useRef(null);
```

We will need to track the page visitor count in the state. We can do this by using a `useState` hook. We will also need to keep a record of the previous location, which we can do using a `useRef` React hook.

6. Update `useEffect` that's related to incoming socket events within `StatsProvider`:

```
React.useEffect(() => {
  if (socket) {
    socket.on("count", (count) => {
      setLiveVisitorCount(count);
    });
    socket.on("page-count", (count) => {
      setPageVisitorCount(count);
    });
  }
}, [socket, setLiveVisitorCount,
  setPageVisitorCount]);
```

When the socket receives a `page-count` event, we update the `pageVisitorCount` value in the state using the `setPageVisitorCount` function.

7. Create a new `useEffect` inside `StatsProvider`:

```
React.useEffect(() => {
    if (socket && previousLocation.current !==
      location.pathname) {
        socket.emit("page-update", {
          currentPage: location.pathname,
          previousPage: previousLocation.current,
        });
        previousLocation.current = location.pathname;
    }
}, [location, socket]);
```

Here is one of the most crucial parts of the code. We add the location to the `useEffect` dependency array so that this code runs whenever the user navigates between a page. Within `useEffect`, we check that the socket is available in the state and that the location update does not match the current location. If both conditions are met, we emit a `page-update` to the server, telling it where we have moved to so that it can keep track of the locations.

8. Update the render of the `StatsProvider.js` file:

```
return (
  <StatsContext.Provider
    value={{
      liveVisitorCount,
      pageVisitorCount,
      connected: socket && socket.connected,
    }}
    {...props}
  />
);
```

By including `pageVisitorCount` in the provider's `value` prop, we can access it via the `useStats` hook in our components.

9. Update `components/VistorCountBadge.js` with the following code:

```
const VisitorCountBadge = () => {
  const { liveVisitorCount, pageVisitorCount,
    connected } = useStats();
  return (
    <p className={'${connected? "bg-blue-200" :"bg-
      red-200"} px-2 py-1 inline-block rounded'}>
      {pageVisitorCount} of {liveVisitorCount} visitors
      on this page
    </p>
  );
};
```

Here, we are retrieving `pageVisitorCount` from the `useStats` hook and rendering it to the screen so that the user can see the value within the badge.

10. Start your Gatsby development server and ensure your socket server is also running. Navigate to `localhost:8000`; you should see the visitor and page count. If you duplicate your browser tab, both numbers should rise, and if you navigate one of these tabs to another page on the site, you should see both tabs' page visitor counts update.

Now that we have implemented an entire feature using sockets, let's summarize what we have learned.

Summary

In this chapter, we learned all about web sockets and how we can use them to utilize real-time data within our Gatsby applications. Then, we implemented a working visitor count that shows the number of people on the current page, as well as the site as a whole. Visitor count statistics is one of a whole host of possible applications for web sockets within a personal site. Perhaps you could take what you have learned here and try and implement article reactions, polls, or even a chat application?

In the next chapter, we will learn about our final advanced concept – localization. We will learn how we can make our Gatsby site support multiple languages for an international audience.

13
Internationalization and Localization

This chapter is all about opening your site up to an international audience. We will talk about patterns you can use to make translating your site as it scales simple! Creating your site in English makes it accessible to the 1.3 billion people in the world who speak the language. However, if we provide users with localization options, we can translate the site into any language, therefore making our site accessible to all.

In this chapter, we will cover the following topics:

- Understanding localization and internationalization
- Implementing routes for internationalization
- Page translations for programmatic pages
- Providing locale translations for single-instance pages

Technical requirements

To navigate this chapter, you will need to have completed *Chapter 12, Using Real-Time Data*.

The code present in this chapter can be found at `https://github.com/PacktPublishing/Elevating-React-Web-Development-with-Gatsby-3/tree/main/Chapter13`.

Understanding localization and internationalization

While the end goal of this chapter is to set up localization, we can make things easier by implementing internationalization first. The terms localization and internationalization are often confused, so let's define these terms properly:

- **Internationalization**: The process of ensuring that your website is created in such a way that it can support different languages, locales, and cultures. Internationalization is all about being proactive in your site's design and development to ensure that you don't have to completely redesign it later when you introduce it to a new market. This could include allowing text to be displayed from right to left as well as left to right, for example.

- **Localization**: Normally conducted after internationalization, localization is the process of adapting your site to meet a new locale requirement. This could be adding a language or cultural requirement.

By spending the time upfront to get the internationalization right within the project, you can save yourself considerable time when you need to add a new locale later down the line. Let's now look at how we can modify our project with an internationalization strategy.

Implementing routes for internationalization

A common approach for large sites to accommodate localization is to prefix all paths with a language code. Let's take our about page, for example – the English (and default) language version of the page is located at /about but the French version of the page might be located at /fr/about and the German version of it at /de/about.

Let's implement this pattern now for our default language of English and add French as a secondary language. We can make this easy with the help of gatsby-theme-i18n:

1. Install the new dependencies:

```
npm install gatsby-theme-i18n gatsby-plugin-react-
helmet react-helmet
```

Here we are installing the gatsby-theme-i18n package and its dependencies. This package automatically creates the route prefixes for us. It also adds language and alternate tags to the head of document. This helps Google identify that two pages contain the same content in different languages.

Important Note

This theme uses `react-helmet`, which may clash with `react-helmet-async`, a package we have used in other chapters. Be sure to check that the head of your document is set up as intended when using both, and if you experience issues stick to a single package.

2. Include the `gatsby-theme-i18n` plugin in your `gatsby-config.js` file:

```
{
  resolve: 'gatsby-theme-i18n',
  options: {
    defaultLang: 'en',
    configPath:
      require.resolve('./i18n/config.json'),
  },
},
```

As part of this configuration, we add a couple of options. `defaultLang` refers to the default language we will use on the site – in this case, this is English, so we use the language code `en`. `configPath` is the configuration path to where we will set up `i18n`. This is most commonly in its own folder, which we will create in the next step.

3. Create a folder in your root directory called `i18n`.

4. Create a new file inside the `i18n` folder called `config.json`, and add the following:

```
[
  {
    "code": "en",
    "hrefLang": "en-US",
    "name": "English",
    "localName": "English",
    "langDir": "ltr",
    "dateFormat": "MM/DD/YYYY"
  },
  {
    "code": "fr",
    "hrefLang": "fr-FR",
```

```
        "name": "French",
        "localName": "Francais",
        "langDir": "ltr",
        "dateFormat": "DD/MM/YYYY"
    }
  ]
```

Here it is important that we define a configuration for every locale we intend to support. Each configuration object must contain the following:

a. `code`: This refers to the language's code you will use to access this locale. Though you can set this to whatever you want for each language, it is probably best to keep them easily identifiable, for example, `fr` for French and `en` for English.

b. `hrefLang`: This is the value that is used for the `hrefLang` tag attribute in the head of the HTML. It is used to tell Google which language you are using on a specific page.

c. `name`: The name of the language in the native language of the developer.

d. `localName`: This is how the name of the language is spelled by its native speakers.

e. `langDir`: This is the direction that text is read in the given locale. This could be `ltr` for left-to-right text or `rtl` for right-to-left text.

f. `dateFormat`: This is the date format used within the locale.

Quick Tip

When adding locales later on, this is the only file that needs to be updated to create the required routes for the locale.

---

5. Finally replace instances of the Gatsby `Link` component on your site with the `LocalizedLink` component from `gatbsy-theme-i18n`:

```
import { LocalizedLink } from "gatsby-theme-i18n";
import React from "react";
const Footer = () => {
  return (
    <footer className="px-2 border-t w-full max-w-5xl
      mx-auto py-4">
      <div className="flex justify-between w-full">
        <div className="flex flex-col justify-center
```

```
                    items-end space-y-1">
                    <p>Your name here.</p>

                    <LocalizedLink to="/" language="en"
                      className="text-blue-400 text-xs">
                      English
                    </LocalizedLink>
                    <LocalizedLink to="/" language="fr"
                        className="text-blue-400 text-xs">
                        Francais
                    </LocalizedLink>
                </div>
            </div>
        </footer>
    );
};
export default Footer;
```

LocalizedLink is a component that extends the Link component with a language prop. By specifying a language's code (from the i18n/config. json file), we route the user to the corresponding page in that specific language. If no language is specified, it will keep the user in their currently active locale. In the preceding example code, we have modified the footer to include links to the index page for both English and French visitors. This will allow site visitors to switch between the locales on any page.

6. Let's verify the previous steps. First, start your Gatsby development server's GraphQL layer (normally located at http://localhost:8000/_graphql) and run the following query:

```
query MyQuery {
    allSitePage {
        nodes {
            path
        }
    }
}
```

In the returned data object, you should be able to see nodes with paths for all locales:

```
{
    "data": {
      "allSitePage": {
        "nodes": [
          {
            "path": "/"
          },
          {
            "path": "/fr/ "
          },
          {
            "path": "/blog"
          },
          {
            "path": "/fr/blog"
          },
          // Continued list
        ]
      }
    },
    "extensions": {}
}
```

7. Finally, navigate to your Gatsby development site. Your footer should now contain links to the two languages:

⊙ London, UK 🔊

Your name here.

English

French

Figure 13.1 – Site footer with language toggle

Clicking **French** should route you to /fr/ and clicking **English** should route you back to /.

Now that we have set up our pages for different locales, let's ensure that we have language-appropriate content to display on these pages. Let's start by looking at pages generated programmatically.

Page translations for programmatic pages

To be able to offer pages such as articles and blog posts translated, we will need to provide the content in both languages. Let's look at how we structure our project so that posts in different languages are available to the site visitor.

`gatsby-theme-i18n` comes with built-in support for handling **MDX** content (a format you can read more about in *Chapter 3, Sourcing and Querying Data (from Anywhere!)*). If you are using Markdown files, this will also work for you. Just ensure that the `gatsby-plugin-mdx` plugin is set up to treat `.md` files as `.mdx` by adding the extension to the plugin's configuration options:

```
{
    resolve: 'gatsby-plugin-mdx',
    options: {
      extensions: ['.mdx', '.md'],
    },
}
```

I will be using the posts that are local files for this demo, but the same steps will work for CMS content once it is ingested to Gatsby's data layer:

1. First, we need to restructure our blog posts folder in a way that makes them easy to identify when we have duplicates in different languages. Instead of filenames, use folder names to group them. Inside `/blog-posts`, create a folder for each post. A good name format for these folders would be `YYYY-mm-DD-Post-Title`. This makes the folder sortable by date but also tells you what the post is about without you having to open the folder.

2. Inside this folder, place the corresponding default language blog post and rename it to `index.mdx`. Be sure that the MDX file contains a `slug` in the frontmatter. An example might look like this:

```
---
type: Blog
title: My First Hackathon Experience
desc: This post is all about my learnings from my
  first hackathon experience in London.
date: 2020-06-20
hero: ../../../assets/images/cover-1.jpeg
tags: [hackathon, webdev, ux]
slug: /my-first-post/
```

```
---
My First Hackathon Experience was great!
Rest of content...
```

3. Repeat *Step 2* for every blog post you wish to add.

4. Create a second file in this folder called `index.fr.mdx`. This file's name has the locale code added between the filename and extension. In this example, we are using France (French), so the locale code is `fr`. Replicate the `index.mdx` file's frontmatter in French by translating all the text values. The `slug`, `type`, `hero`, `date`, and `tags` must remain the same for both files. The resultant file for the example started in *Step 2* looks as follows:

```
---
type: Blog
title: Ma première expérience de hackathon
desc: Ce post est tout sur mes apprentissages de ma
   première expérience de hackathon à Londres.
date: 2020-06-20
hero: ../../../assets/images/cover-1.jpeg
tags: [hackathon, webdev, ux]
slug: /my-first-post/
---
Ma première expérience de hackathon était super !
Rest of content...
```

5. Repeat *Step 4* for any additional languages and locales you wish to support.

6. Update your `gatsby-node.js` files' blog post page creation configuration with the following:

```
exports.createPages = async ({ actions, graphql,
  reporter }) => {
  const { createPage } = actions;
  const BlogPostTemplate =
    path.resolve('./src/templates/blog-page.js');
  const BlogPostQuery = await graphql('
    {
      allMdx(filter: { frontmatter: { type: { eq:
        "Blog" } } }) {
```

```
              nodes {
                frontmatter {
                  slug
                }
              }
            }
          }
  ');
  const BlogPosts = BlogPostQuery.data.allMdx.nodes;
  BlogPosts.forEach(({ frontmatter: { slug } }) => {
    createPage({
      path: slug,
      component: BlogPostTemplate,
      context: {
        slug: slug,
      },
    });
  });
};
```

Here, we query for the `slug` from the frontmatter of the MDX files. We then use this to create the page with the `createPage` function, ensuring that we also provide the `slug` to the component as `context`. `atsby-theme-i18n` listens for page creation and will additionally create the same page for each locale without any additional configuration! It will also add two fields to the MDX nodes in our GraphQL data layer – `locale` and `isDefault`, which tell you what locale the MDX is and whether the MDX is the default locale, respectively.

7. We now need to tell Gatsby to use the correct MDX file on the correct locale path. Without this step, your site will find the MDX file with the first matching `slug` when creating blog posts. This may not match the locale as we have multiple files with the same `slug` and could lead to us being lost in translation. First, update the blog page template (located at `src/templates/blog-page.js`) query with the following:

```
export const pageQuery = graphql'
  query($locale: String!, $slug: String!) {
    blogpost: mdx(
      frontmatter: { slug: { eq: $slug } }
```

```
      fields: { locale: { eq: $locale } }
  ) {
    frontmatter {
      date
      title
      desc
      tags
      hero {
        childImageSharp {
          gatsbyImageData(width: 600, height: 400,
            placeholder: BLURRED)
        }
      }
    }
    body
  }
}
';
```

Here, we access the locale field provided via the gatsby-theme-i18n plugin and use it to filter the MDX blog posts to those that match the specified locale. This will ensure that we render the blog post in the correct language on any blog page.

8. Perform the exact same step in the src/templates/blog-preview.js file:

```
export const pageQuery = graphql'
  query($locale: String!,$skip: Int!, $limit: Int!) {
    blogposts: allMdx(
      limit: $limit
      skip: $skip
      filter: {frontmatter: {type: {eq: "Blog"}},
      fields: {locale: { eq: $locale }}}
      sort: { fields: frontmatter___date, order: DESC }
    ) {
      nodes {
        frontmatter {
          date
          title
```

```
            tags
            desc
            slug
            hero {
              childImageSharp {
                gatsbyImageData(width: 240, height: 160,
                  placeholder: BLURRED)
              }
            }
          }
        }
      }
    }
  ';
```

9. Let's verify the previous steps. First, start your Gatsby development server's GraphQL layer (normally located at `http://localhost:8000/_graphql`) and run the following query:

```
query MyQuery {
  blogposts: allMdx(filter: {frontmatter: {type: {eq:
    "Blog"}}}) {
    nodes {
      fields {
        locale
        isDefault
      }
      frontmatter {
        slug
      }
    }
  }
}
```

Here, we are querying for all MDX of type `Blog` and retrieving the locale, whether that locale is the default, and the `slug`. The result should look like this:

```
{
    "data": {
        "blogposts": {
            "nodes": [
                {
                    "fields": {
                        "locale": "en",
                        "isDefault": true
                    },
                    "frontmatter": {
                        "slug": "/my-first-post/"
                    }
                },
                {
                    "fields": {
                        "locale": "fr",
                        "isDefault": false
                    },
                    "frontmatter": {
                        "slug": "/my-first-post/"
                    }
                }
            ]
        }
    }
}
```

A result should be present for each locale for any given `slug`. Assuming this is the case, you can navigate to `/blog` on your Gatsby development site. You should see your blog content in your default language:

An Introduction to Gatsby

2020-07-22 `GATSBY` `WEBDEV`

Gatsby is an awesome tool - here is how to get started.

My First Hackathon Experience

2020-06-20 `HACKATHON` `WEBDEV` `UX`

This post is all about my learnings from my first hackathon experience in London.

Figure 13.2 – Blog page in English

Navigating to /fr/blog, you should see your content in French:

Une introduction à Gatsby

2020-07-22 `GATSBY` `WEBDEV`

Gatsby est un outil génial - voici comment commencer.

Ma première expérience de hackathon

2020-06-20 `HACKATHON` `WEBDEV` `UX`

Ce post est tout sur mes apprentissages de ma première expérience de hackathon à Londres.

Figure 13.3 – Blog page in French

Quick Tip

If you are clicking on a blog post and always receiving the default locale version, the most likely cause is that you are using `Link` instead of `LocalizedLink` when navigating to the page. Review *Step 5* of the *Implementing routes for internationalization* section of this chapter.

We can provide translations for our programmatically generated pages with ease using this strategy. Let's now learn how we can set up translations for single-instance pages.

Providing locale translations for single-instance pages

For static pages, we will need a different approach to providing translations. For any strings that require translation we can no longer have the values in line with our JSX. A very common approach is to use `react-i18next`, which has a great hook called `useTranslation` that allows you to switch strings out depending on the locale. Let's use this now to translate content on our index page for site visitors:

1. Open a terminal at your root directory and add these new dependencies:

    ```
    npm install gatsby-theme-i18n-react-i18next react-
    i18next i18next
    ```

    Here, we are installing the `gatsby-theme-i18n-react-i18next` package and its dependencies. This package is a Gatsby theme plugin that provides locale support to our application by wrapping our site in `react-i18next`'s context provider. Underneath the hood, this package wraps the site by using `wrapPageElement` in the `gatsby-browser.js` in the same way we did in *Chapter 12, Using Real-Time Data*.

2. Include the `gatsby-theme-i18n-react-i18next` plugin in your `gatsby-config.js` file:

    ```
    {
        resolve: 'gatsby-theme-i18n-react-i18next',
        options: {
          locales: './i18n/locales',
          i18nextOptions: {
            ns: ["globals"],
          },
        },
    },
    ```

    As part of this configuration, we add a couple of options. `locales` refer to location where we will store our translations. `i18nextOptions` accepts any configuration options that `i18next` accepts (the full list is available at `https://www.i18next.com/overview/configuration-options`). Here, we are passing in the `ns` option, which is an array of namespaces to use. For this example, we will just be creating a single namespace called `globals`, but you may want to add more as your site grows.

> **Important Note**
> The `gatsby-theme-i18n-react-18next` is an add-on package that will only work in tandem with `gatsby-theme-i18n`. Ensure that this package is installed by following the steps in the *Implementing routes for internationalization* section.

3. Create a new folder in `i18n` called `locales`.

4. Within `locales`, create a new folder for each locale your site supports, for example, `en` and `fr`.

5. For each namespace, create a JSON file in the `locale` folder. In our case, we need to create a single file named `globals.json` for our `globals` namespace in each folder. This file should contain any translations you require, retrievable with a key that is consistent across all files. Your English file (which should be located at `i18n/locales/en/globals.json`) should contain the following:

```
{
    "header": "Site Header",
    "yourName": "Your Name",
    "aboutMe": "About Me",
    "location": "London, UK",
    "bio": "A short biography about me"
}
```

Your French file (which should be located at `i18n/locales/fr/globals.json`) should contain the following:

```
{
    "header": "En-tête du site",
    "yourName": "Votre nom",
    "aboutMe": "À propos de moi",
    "location": "France, Paris",
    "bio": "Une courte biographie sur moi."
}
```

6. To use the translation within a Gatsby page component, we can use the useTranslation hook from react-i18next. Let's look at the about me link on the index page (located at src/pages/index.js) as an example:

```
import React from "react";
import { useTranslation } from "react-i18next";
import { LocalizedLink } from "gatsby-theme-i18n";
import Layout from "../components/layout/Layout";
import SEO from "../components/layout/SEO";

export default function SamplePage() {
  const { t } = useTranslation("globals");
  return (
    <Layout>
      <SEO title="Home" description="The landing page
        of my website" />
      <div className="max-w-5xl mx-auto py-16 lg:py-
        24">
        <LocalizedLink to="/about" className="btn">
          {t("aboutMe")}
        </LocalizedLink>
      </div>
    </Layout>
  );
}
```

We import useTranslation from reacti18next. Then within the page component, we invoke the hook specifying the namespace we wish to use. In our case, this is the globals namespace we have created. The t function can be used to retrieve the translation from the namespace by passing in a valid key from the globals.json objects created in *Step 5*. t("aboutMe") will return *About me* when on the en locale and *À propos de moi* when on the fr locale.

7. We can also use the exact same process in any other components, such as our header, for example:

```
import React from "react";
import { useTranslation } from "react-i18next";
import { LocalizedLink } from "gatsby-theme-i18n"
```

```
const Header = () => {
  const { t } = useTranslation("globals");
  return(
  <header className="px-2 border-b w-full max-w-7xl
    mx-auto py-4 flex items-center justify-between">
    <LocalizedLink to="/">
      <div className="flex items-center space-x-2
        hover:text-blue-600">
        <p className="font-bold text-
          2xl">{t("header")}</p>
      </div>
    </LocalizedLink>
  </header>
  )
};
export default Header;
```

You can even use this inside React components that are used within MDX content if you need to!

8. Verify your implementation by navigating to the index page on your Gatsby development site. Toggle the locale by modifying the path or using the Footer component, and you should see any copy that is using useTranslation update.

We've only scratched the surface of the features that i18next offers. Visit their documentation at https://www.i18next.com/ to learn more about the powerful capabilities they offer. With this strategy and the preceding sections, you should now feel confident translating any aspect of your site. Let's now summarize what we've learned.

Summary

In this final chapter, we learned about making our site accessible to a global audience. We first identified the differences between internationalization and localization. We then used the gatsby-theme-i18n plugin to create routes for our locales. We created programmatic blog posts in different languages and ensured the correct translation was visible when visiting a locale. Finally, we also translated our static pages using the gatsby-theme-i18n-react-i18next plugin. Between these two plugins, you now have the power to translate any of your site's content.

Hi!

I am Samuel Larsen-Disney, author of Elevating React Web Development with Gatsby. I really hope you enjoyed reading this book and found it useful for increasing your productivity and efficiency in Gatsby.

It would really help me (and other potential readers!) if you could leave a review on Amazon sharing your thoughts on Elevating React Web Development with Gatsby here.

Go to the link below or scan the QR code to leave your review:

`https://packt.link/r/1800209096`

Your review will help me to understand what's worked well in this book, and what could be improved upon for future editions, so it really is appreciated.

Best Wishes,

Samuel Larsen-Disney

Index

Packt.com

Subscribe to our online digital library for full access to over 7,000 books and videos, as well as industry leading tools to help you plan your personal development and advance your career. For more information, please visit our website.

Why subscribe?

- Spend less time learning and more time coding with practical eBooks and Videos from over 4,000 industry professionals

- Improve your learning with Skill Plans built especially for you

- Get a free eBook or video every month

- Fully searchable for easy access to vital information

- Copy and paste, print, and bookmark content

Did you know that Packt offers eBook versions of every book published, with PDF and ePub files available? You can upgrade to the eBook version at packt.com and as a print book customer, you are entitled to a discount on the eBook copy. Get in touch with us at customercare@packtpub.com for more details.

At www.packt.com, you can also read a collection of free technical articles, sign up for a range of free newsletters, and receive exclusive discounts and offers on Packt books and eBooks.

Other Books You May Enjoy

If you enjoyed this book, you may be interested in these other books by Packt:

Full-Stack React Projects

Shama Hoque

ISBN: 978-1-83921-541-4

- Extend a basic MERN-based application to build a variety of applications
- Add real-time communication capabilities with Socket.IO
- Implement data visualization features for React applications using Victory
- Develop media streaming applications using MongoDB GridFS
- Improve SEO for your MERN apps by implementing server-side rendering with data
- Implement user authentication and authorization using JSON web tokens
- Set up and use React 360 to develop user interfaces with VR capabilities
- Make your MERN stack applications reliable and scalable with industry best practices

React Projects

Roy Derks

ISBN: 978-1-80107-063-8

- Create a wide range of applications using various modern React tools and frameworks
- Discover how React Hooks modernize state management for React apps
- Develop web applications using styled and reusable React components
- Build test-driven React applications using Jest, React Testing Library, and Cypress
- Understand full-stack development using GraphQL, Apollo, and React
- Perform server-side rendering using React and Next.js
- Create animated games using React Native and Expo
- Design gestures and animations for a cross-platform game using React Native

Packt is searching for authors like you

If you're interested in becoming an author for Packt, please visit authors. packtpub.com and apply today. We have worked with thousands of developers and tech professionals, just like you, to help them share their insight with the global tech community. You can make a general application, apply for a specific hot topic that we are recruiting an author for, or submit your own idea.

www.ingramcontent.com/pod-product-compliance
Lightning Source LLC
Chambersburg PA
CBHW080929060326
40690CB00042B/3240

9781800209091